Compact Textbooks in Mathematics

This textbook series presents concise introductions to current topics in mathematics and mainly addresses advanced undergraduates and master students. The concept is to offer small books covering subject matter equivalent to 2- or 3-hour lectures or seminars which are also suitable for self-study. The books provide students and teachers with new perspectives and novel approaches. They may feature examples and exercises to illustrate key concepts and applications of the theoretical contents. The series also includes textbooks specifically speaking to the needs of students from other disciplines such as physics, computer science, engineering, life sciences, finance.

- **compact:** small books presenting the relevant knowledge
- **learning made easy:** examples and exercises illustrate the application of the contents
- **useful for lecturers:** each title can serve as basis and guideline for a semester course/lecture/seminar of 2-3 hours per week.

Alberto Facchini

Introduction to Ring and Module Theory

 Birkhäuser

Alberto Facchini
Dipartimento di Matematica "Tullio Levi-Civita"
Università di Padova
Padova, Italy

ISSN 2296-4568   ISSN 2296-455X   (electronic)
Compact Textbooks in Mathematics
ISBN 978-3-031-82508-8   ISBN 978-3-031-82509-5   (eBook)
https://doi.org/10.1007/978-3-031-82509-5

© The Editor(s) (if applicable) and The Author(s), under exclusive license to Springer Nature Switzerland AG 2025

This work is subject to copyright. All rights are solely and exclusively licensed by the Publisher, whether the whole or part of the material is concerned, specifically the rights of translation, reprinting, reuse of illustrations, recitation, broadcasting, reproduction on microfilms or in any other physical way, and transmission or information storage and retrieval, electronic adaptation, computer software, or by similar or dissimilar methodology now known or hereafter developed.
The use of general descriptive names, registered names, trademarks, service marks, etc. in this publication does not imply, even in the absence of a specific statement, that such names are exempt from the relevant protective laws and regulations and therefore free for general use.
The publisher, the authors and the editors are safe to assume that the advice and information in this book are believed to be true and accurate at the date of publication. Neither the publisher nor the authors or the editors give a warranty, expressed or implied, with respect to the material contained herein or for any errors or omissions that may have been made. The publisher remains neutral with regard to jurisdictional claims in published maps and institutional affiliations.

This book is published under the imprint Birkhäuser, www.birkhauser-science.com by the registered company Springer Nature Switzerland AG
The registered company address is: Gewerbestrasse 11, 6330 Cham, Switzerland

If disposing of this product, please recycle the paper.

Preface

This textbook is the culmination of my experiences teaching the course of "Introduction to Ring Theory" at the University of Padua over the past 15 years. Initially, I relied on other textbooks, but I found them either too heavy on introductory notions or too replete with examples, hindering the coverage of essential results like the Artin-Wedderburn theorem. Consequently, I began crafting and refining my own lecture notes, which have evolved into this book.

As far as prerequisites are concerned, in order to follow this textbook students must be familiar with the algebraic concepts studied in the first two years of a mathematics degree: linear algebra, basic group theory, principal integral domains, unique factorization domains, and Euclidean domains. Of course, additional knowledge in areas such as Galois theory, commutative algebra, algebraic geometry, or group theory will be advantageous. The selection of topics presented here stems from my personal views on what is essential for students at this stage (last year undergraduate students in mathematics and master students) relatively to their mathematical education. I believe the topics I have chosen are crucial for the students' mathematical growth, their understanding of algebra's foundational ideas, and their development of an algebraic mindset.

Over the years, student interests have varied, prompting me to expand and adapt the material. While the full breadth of this textbook now exceeds what can be covered in a 60-hour course, this provides flexibility for lecturers to focus on specific areas. Numerous exercises are included to encourage active learning and problem-solving. The exercises proposed are fundamental in the learning of the topics presented.

The textbook integrates elementary category theory, with basic concepts and examples developed throughout the course. Although the primary focus is on (noncommutative) rings and modules, relevant notions for other algebraic structures, such as groups and semigroups, are also discussed. Thus, this book aims to introduce students to noncommutative rings and modules within a broader algebraic context. Here is an overview of the chapters. Chapter 1 introduces foundational concepts, including the definitions of rings, categories, functors, modules, and sets of generators. Chapter 2 is devoted to some key classes of modules: the classes of free, projective, simple, semisimple, Noetherian, Artinian modules, and the class of modules of finite composition length. Chapter 3 focuses on Artinian rings, presenting significant results such as the Artin-Wedderburn theorem, the Hopkins-

Levitzki theorem, the study of the Jacobson radical, and Maschke's theorem. Chapter 4 delves into other fundamental ring theory notions: local rings, injective modules, projective covers, injective envelopes, Goldie dimension, direct limits, and inverse limits. Chapter 5 introduces additive categories and abelian categories.

In my view, the first 110 pages should be covered in any introductory ring and module theory course. The remaining content can be tailored to the lecturer's preferences and the students' interests. After four decades of teaching algebra, I believe these topics are vital for students to grasp the core concepts of algebra, operations, and algebraic structures.

Over the years, students have appreciated this selection of topics and the way they are presented. I am deeply grateful for the enthusiasm and satisfaction they have shown, which has continually inspired and motivated me.

Padua, Italy   Alberto Facchini
June 2024

Contents

1 Basic Notions 1
   1.1   Rings 1
   1.2   Categories 7
   1.3   Functors 10
   1.4   Modules 12
   1.5   Bimodules 19
   1.6   Submodules and Quotients 22
   1.7   Natural Transformations of Functors 25
   1.8   Sums and Intersections, Direct Sums and Direct Products 30
   1.9   Isomorphism Theorems 34
   1.10 Sets of Generators, Maximal Submodules 36
   1.11 The Lattice of Submodules 38

2 Some Classes of Modules 41
   2.1   Free Modules 41
   2.2   Further Properties of Free Modules 49
   2.3   Further Terminology About Categories 54
   2.4   IBN Rings 59
   2.5   Exact Sequences 61
   2.6   Projective Modules 64
   2.7   Tensor Product of Modules 67
   2.8   Projective $\mathbb{Z}$-Modules: Hereditary Rings 76
   2.9   Idempotents, Nilpotents, Further Results on Projective Modules ... 78
   2.10 Simple Modules, Semisimple Modules 82
   2.11 Noetherian Modules, Artinian Modules 87
   2.12 Modules Over the Factor Ring R/I 91
   2.13 Series of Modules: Modules of Finite Composition Length 93

3 Right Artinian Rings 101
   3.1   Semisimple Artinian Rings 101
   3.2   The Artin-Wedderburn Theorem 107
   3.3   The Nilradical, Right Artinian Rings, and the
         Hopkins-Levitzki Theorem 111
   3.4   The Radical of a Module 115

	3.5	The Jacobson Radical of a Ring	117
	3.6	Group Representations	121
4	**Local Rings, Injective Modules, Flat Modules**		129
	4.1	Local Rings	129
	4.2	Injective Modules	136
	4.3	Projective Covers	142
	4.4	Injective Envelopes	144
	4.5	Uniform Modules, Goldie Dimension	150
	4.6	Direct Limits of Modules	157
	4.7	Direct Limits of Rings	163
	4.8	Flat Modules	164
	4.9	Finitely Presented Modules	167
	4.10	Inverse Limits	170
5	**Additive Categories, Abelian Categories**		173
	5.1	Monomorphisms, Epimorphisms, Subobjects, Quotient Objects ...	173
	5.2	Products and Coproducts	174
	5.3	Initial Objects, Preadditive Categories	177
	5.4	Additive Categories	182
	5.5	Equalizers, Coequalizers, Kernels and Cokernels	184
	5.6	Abelian Categories	186
6	**Appendices**		193
	6.1	Appendix 1: Axiomatic Set Theory	193
	6.2	Appendix 2: Cardinal Numbers, Ordinal Numbers	197

For Further Study... 215

Bibliography 217

Index 219

Basic Notions

1.1 Rings

In these notes, all the rings we will consider will be associative rings with identity. We will mostly denote our rings by R or, when we want to be more precise, by a triple $(R, +, \cdot)$. Thus R will be a set and $+, \cdot$ will be two binary operations, such that

(i) $(R, +)$ is an abelian group:
   (a) $a + (b + c) = (a + b) + c$ for every $a, b, c \in R$;
   (b) $a + b = b + a$ for every $a, b \in R$;
   (c) there exists an element 0_R of R such that $a + 0_R = a$ for every $a \in R$;
   (d) for every $a \in R$ there is an element $b \in R$ such that $a + b = 0_R$ (This element b is then necessarily unique and usually denoted $-a$.)
(ii) $(R, \cdot)$ is a semigroup with identity, i.e., a monoid:
   (e) $a(bc) = (ab)c$ for every $a, b, c \in R$;
   (f) there exists an element 1_R of R such that $a \cdot 1_R = 1_R \cdot a = a$ for every $a \in R$.
(iii) Multiplication is both right and left distributive over addition:
   (g) $a(b + c) = ab + ac$ for every $a, b, c \in R$;
   (h) $(a + b)c = ac + bc$ for every $a, b, c \in R$.

A ring R is *commutative* if the multiplication $\cdot$ is commutative: $ab = ba$ for every $a, b \in R$.

Exercise 1.1.1
For every element a of a ring R, one has $a \cdot 0_R = 0_R \cdot a = 0_R$.

▶ **Remarks 1.1.2**

(1) Let R be a ring in which the identity 0_R with respect to the operation $+$ coincides with the identity 1_R with respect to the operation $\cdot$. Then, for every $a \in R$, we have that
$$a = a \cdot 1_R = a \cdot 0_R = 0_R.$$
In other words, R is the set with one element:
$$R = \{0_R\}.$$
This ring R is called the *zero ring*, denoted by $R = 0$.

(2) Consider a "ring without identity", that is a triple $(R, +, \cdot)$, where R is a set and $+, \cdot$ are two binary operations, for which all the previous axioms of rings hold, except for (f), so that $(R, \cdot)$ is a semigroup but not a monoid. Consider the direct sum $R \oplus \mathbb{Z}$ of the additive group of R and the additive group $\mathbb{Z}$ of integers. Define on the abelian group $R \oplus \mathbb{Z}$ a multiplication by
$$(r, z) \cdot (r', z') = (rr' + zr' + z'r, zz') \qquad \text{for every } (r, z), (r', z') \in R \oplus \mathbb{Z}.$$
Here zr' and $z'r$ denote the z-th multiple of r' and the z'-th multiple of r in the additive group of R, respectively. Then $R \oplus \mathbb{Z}$ turns out to be a ring with identity $(0_R, 1_\mathbb{Z})$. Notice that this identity is different from the zero $(0_R, 0_\mathbb{Z})$ of the ring $R \oplus \mathbb{Z}$. Moreover, there is a canonical embedding $\varepsilon \colon R \to R \oplus \mathbb{Z}$, $\varepsilon(r) = (r, 0_\mathbb{Z})$, that is, an injective mapping that preserves the two operations $+$ and $\cdot$. This shows that, "enlarging" R, we have been able to embed R into a ring with identity.

A *subring* of a ring R is a subset S of R that is a subgroup of $(R, +)$ and a submonoid of $(R, \cdot)$. That is,

(i) $S \subseteq R$;
(ii) $a - b \in S$ for every $a, b \in S$;
(iii) $ab \in S$ for every $a, b \in S$;
(iv) $1_R \in S$.

A *right ideal* I of R is a subgroup of $(R, +)$ such that $xr \in I$ for every $x \in I$ and every $r \in R$. That is,

(i) $I \subseteq R$;
(ii) $I \neq \emptyset$;
(iii) $x - y \in I$ for every $x, y \in I$;
(iv) $xr \in I$ for every $x \in I, r \in R$.

We will denote "I is a right ideal of R" by $I \leq R_R$.

1.1 Rings

A *left ideal* I of R is a subgroup of $(R, +)$ such that $rx \in I$ for every $r \in R$ and every $x \in I$. We will denote "I is a left ideal of R" by $I \leq {}_R R$.

A *two-sided ideal R* (or, simply, an *ideal* of R) is a subset I of R that is both a right ideal of R and a left ideal of R. We will write $I \trianglelefteq R$. A subring, right ideal, left ideal or two-sided ideal of R is *proper* if it is different from R.

If I is a two-sided ideal of R, the *quotient ring R/I of R modulo I* is the set of all cosets $r + I$, where r ranges in R, and with the operations defined by $(r + I) + (s + I) = (r+s) + I$, $(r + I)(s + I) = (rs) + I$. These are well defined operations. The identity is $1_{R/I} = 1_R + I$.

A *ring morphism* (or *ring homomorphism*) $\varphi \colon R \to S$ is a mapping between two rings R and S such that, for every $a, b \in R$, we have that $\varphi(a + b) = \varphi(a) + \varphi(b)$, $\varphi(ab) = \varphi(a)\varphi(b)$, and $\varphi(1_R) = 1_S$.

If $\mathbb{Z}$ is the ring of integers, there is a unique ring morphism $\lambda \colon \mathbb{Z} \to R$. To see this, let us see the existence first and then the uniqueness. In order to prove the existence, define $\lambda \colon \mathbb{Z} \to R$ setting $\lambda(z) = z1_R$ for every $z \in \mathbb{Z}$. Here $z1_R$ denotes the z-th multiple of the element 1_R of the additive group R. It is easily seen that the mapping λ is a ring morphism. To prove the uniqueness, let $\varphi \colon \mathbb{Z} \to R$ be a ring morphism. Then $\varphi(1) = 1_R$ (according to our definitions, ring homomorphisms send the identity to the identity), so that $\varphi(z \cdot 1) = z\varphi(1) = z1_R$ (group morphisms respect multiples). Thus $\varphi(z) = \varphi(z \cdot 1) = z1_R = \lambda(z)$ for every z, hence $\varphi = \lambda$. This shows that λ is unique.

If $\varphi \colon R \to S$ is a ring morphism, the *kernel* of φ is

$$\ker \varphi := \{ r \in R \mid \varphi(r) = 0_S \}.$$

The kernel of a morphism $\varphi \colon R \to S$ is a two-sided ideal of R.

Exercise 1.1.3
Let R be a ring without identity as defined in Remark 1.1.2(2). Prove that the construction of the ring $R \oplus \mathbb{Z}$ and the injective mapping $\varepsilon \colon R \to R \oplus \mathbb{Z}$ preserving the two operations $+$ and $\cdot$ satisfy the following *Universal Property*. For every ring (with identity) S and every mapping $\varphi \colon R \to S$ that preserves the two operations $+$ and $\cdot$, there is a unique ring homomorphism $\varphi' \colon R \oplus \mathbb{Z} \to S$ such that $\varphi = \varphi' \circ \varepsilon$.

Notice that if the ring R has an identity 1_R, then the injective mapping $\varepsilon \colon R \to R \oplus \mathbb{Z}$ does not map the identity 1_R of R to the identity $(0_R, 1_\mathbb{Z})$ of $R \oplus \mathbb{Z}$.

A ring R is *simple* if it is non-zero and the only two-sided ideals of R are $\{0_R\}$ and R.

Exercise 1.1.4
Show that if R is a simple ring and S is a non-zero ring, then every morphism $\varphi \colon R \to S$ is an injective mapping. [Hint: The kernel must be a proper ideal of R.]

An element r of a ring R is said to be:

(i) a *right zero-divisor* if $r \neq 0$ and there exists $s \in R$ such that $s \neq 0$ and $sr = 0$;
(ii) a *left zero-divisor* if $r \neq 0$ and there exists $s \in R$ such that $s \neq 0$ and $rs = 0$;
(iii) a *zero-divisor* if it is either a right zero-divisor or a left zero-divisor;
(iv) *right invertible* if there exists $s \in R$ such that $rs = 1_R$;
(v) *left invertible* if there exists $s \in R$ such that $sr = 1_R$;
(vi) *invertible* if it is both right invertible and left invertible.

A ring R is an *integral domain* if for any integer $n \geq 0$ and $r_1, r_2, \ldots, r_n$ non-zero elements of R, the product $r_1 r_2 \cdots r_n$ is non-zero. For $n = 0$, this conditions says that the product $r_1 r_2 \cdots r_n$ of an empty family of elements of R is non-zero, that is $1_R \neq 0_R$. Hence, equivalently, a ring R is an integral domain if and only if $R \neq 0$ and the product of two non-zero elements of R is non-zero.

▶ **Remark 1.1.5** By definition, if $r \in R$ is invertible, then it is *both* right invertible *and* left invertible. This means that there exist elements $s \in R$ such that $rs = 1_R$ and $s' \in R$ such that $s'r = 1_R$. But then $s' = s' \cdot 1_R = s'(rs) = (s'r)s = 1_R \cdot s = s$. It follows that an invertible element r has a unique *right inverse*, a unique *left inverse*, and the unique right inverse is equal to the unique left inverse. This unique right/left inverse of r will be denoted by r^{-1}.

Exercise 1.1.6
Show that, in a ring $R \neq 0$, every left invertible element is different from zero. Similarly for right invertible elements. Show that left invertible elements are not left zero-divisors, and right invertible elements are not right zero-divisors.

Exercise 1.1.7
Let $R \neq 0$ be a finite ring, that is, a ring with finitely many elements. Show that the following conditions are equivalent for an element $r \in R$:
(i) r is left invertible.
(ii) r is not a left zero-divisor and $r \neq 0$.
(iii) r is right invertible.
(iv) r is not a right zero-divisor and $r \neq 0$.

Exercise 1.1.8
Let k be a field and V, W vector spaces over k. Recall that a mapping $f: V \to W$ is called k-*linear* or a *linear transformation* if $f(v+v') = f(v) + f(v')$ and $f(\lambda v) = \lambda f(v)$ for every $v, v' \in V$ and every $\lambda \in k$. Let $\text{End}_k(V)$ be the set of all k-linear mappings $f: V \to V$. Define two operations on $\text{End}_k(V)$ setting

$$(f+g)(v) := f(v) + g(v)$$
$$(fg)(v) := f(g(v))$$

for every $f, g \in \text{End}_k(V)$ and every $v \in V$. Then $\text{End}_k(V)$ turns out to be a ring in which the multiplication is the composition of linear mappings, the identity $1_{\text{End}_k(V)}$ is the identity mapping $V \to V$, $v \mapsto v$ for every $v \in V$, and the zero $0 = 0_{\text{End}_k(V)}$ is the zero mapping, that is,

1.1 Rings

the constant mapping $V \to V$, $v \mapsto 0_V$ for every $v \in V$. The ring $\text{End}_k(V)$ is called the *endomorphism ring* of the vector space k.

(1) Show that the following conditions are equivalent for a non-zero element $f \in \text{End}_k(V)$:
   (i) f is an injective mapping;
   (ii) f is left invertible in the ring $\text{End}_k(V)$;
   (iii) f is not a left zero-divisor.
(2) Show that the following conditions are equivalent for a non-zero element $f \in \text{End}_k(V)$:
   (i) f is a surjective mapping;
   (ii) f is right invertible in the ring $\text{End}_k(V)$;
   (iii) f is not a right zero-divisor.
(3) Show that an element $f \in \text{End}_k(V)$ is a zero-divisor if and only if $f \neq 0$ and f is not a bijection.
(4) Show that an element $f \in \text{End}_k(V)$ is invertible if and only if it is a bijection.
(5) Give examples of elements in $\text{End}_k(V)$ that have different *left inverses* (but no right inverses), and examples of elements in $\text{End}_k(V)$ that have different right inverses (but no left inverses). Notice that for $V = 0$, the ring $\text{End}_k(V)$ is the zero ring.

Exercise 1.1.9
Let M be an additive abelian group. Let $\text{End}(M)$ be the set of all endomorphisms of the group M. For every $\varphi, \psi \in \text{End}(M)$, define $(\varphi + \psi)(x) = \varphi(x) + \psi(x)$ and $(\varphi\psi)(x) = \varphi(\psi(x))$, $x \in M$. Prove that $\text{End}(M)$ is a ring with respect to these two operations. It is called the *endomorphism ring* of the abelian group M. Determine the identity and the zero element of $\text{End}(M)$.

A ring R is a *division ring* if its invertible elements are exactly its non-zero elements. Notice that in the zero ring the zero element is invertible. Hence the zero ring is not a division ring.

> **Lemma 1.1.10**
> *The following conditions are equivalent for a ring $R \neq 0$:*
>
> (i) *R is a division ring.*
> (ii) *Every non-zero element of R is right invertible.*
> (iii) *Every non-zero element of R is left invertible.*
> (iv) *The only right ideals of R are $\{0_R\}$ and R.*
> (v) *The only left ideals of R are $\{0_R\}$ and R.*

Proof
(i) $\Rightarrow$ (ii)   is trivial.
(ii) $\Rightarrow$ (i)   Let R be a ring in which every non-zero element is right invertible. We must show that every non-zero element x of R is invertible. Now if x is a non-zero element, x is right invertible, hence there exists $y \in R$ with $xy = 1$. The element y is non-zero, otherwise $1 = xy = x \cdot 0 = 0$. Now y non-zero implies y right invertible, so that there exists $z \in R$ with $yz = 1$. Therefore x is a left

inverse of y and z is a right inverse of y. It follows that $x = z$ (Remark 1.1.5), so that $yx = yz = 1$, that is, y is a left inverse of x as well. This shows that x is invertible.

(ii) $\Rightarrow$ (iv)   Assume that (ii) holds. Let I be a non-zero right ideal of R. Let $x \in I$ be a non-zero element. Then x is right invertible, so that there exists $y \in R$ with $xy = 1$. Then $I = R$, because for every $r \in R$ we have that $r = 1 \cdot r = xyr \in I$.

(iv) $\Rightarrow$ (ii)   Assume that (iv) holds. Let x be a non-zero element of R. It is easily checked that the set $xR := \{xr \mid r \in R\}$ is a right ideal of R. Notice that $x = x \cdot 1$ belong to xR because R has identity 1. Thus $x \neq 0$ implies that xR is a non-zero right ideal. By (iv), $xR = R$. Thus there exists $t \in R$ with $1 = xt$, so that x is right invertible.

Similarly, (i) $\Leftrightarrow$ (iii) and (iii) $\Leftrightarrow$ (v).

$\square$

Exercise 1.1.11

For every ring R and integer $n \geq 1$, let $\mathbb{M}_n(R)$ be the ring of all $n \times n$ matrices with entries in R. Here the addition is componentwise, and the multiplication is the matrix multiplication (row by column). Show that if k is a division ring, then the ring $\mathbb{M}_n(k)$ is simple. Show that $\mathbb{M}_n(k)$ is not a division ring for $n \geq 2$. Thus every division ring is a simple ring, but not conversely.

[Sketch of the solution: Assume that k is a division ring. Let I be a non-zero two-sided ideal of the ring $\mathbb{M}_n(k)$. Let $A = (a_{i,j})_{i,j}$ be a non-zero element of I. Then there exist two indices $\bar{i}, \bar{j}$ with $a_{\bar{i},\bar{j}} \neq 0$. Thus in I we can find the matrix $E_{\bar{i},\bar{i}} A E_{\bar{j},\bar{j}}$, which is a matrix in which all entries are 0 except for the $(\bar{i}, \bar{j})$-entry, which is equal to $a_{\bar{i},\bar{j}}$. Multiplying this matrix in I with the scalar matrix
$\begin{pmatrix} \lambda a_{\bar{i},\bar{j}}^{-1} & 0 & \cdots & 0 \\ 0 & \lambda a_{\bar{i},\bar{j}}^{-1} & & 0 \\ \vdots & & \ddots & \\ 0 & & & \lambda a_{\bar{i},\bar{j}}^{-1} \end{pmatrix}$, with $\lambda \in k$, we find that in I there are all the matrices A in
which all entries are 0 except for the $(\bar{i}, \bar{j})$-entry, which can be any element λ of k. Multiplying these elements of I by $E_{i,\bar{i}}$ on the left and by $E_{\bar{j},j}$ on the right, we obtain that in I there are all the matrices $E_{i,\bar{i}} A E_{\bar{j},j}$, which are the matrices in which all entries are 0 except for the (i, j)-entry, which is λ. It is clear that any matrix in $\mathbb{M}_n(k)$ is a sum of at most n^2 matrices of this type of matrices in I, so that every matrix in $\mathbb{M}_n(k)$ belongs to I. It follows that $I = \mathbb{M}_n(k)$.]

Exercise 1.1.12

For every ring R and integer $n \geq 1$, show that there is a one-to-one correspondence between the set $\mathcal{I}(R)$ of all two sided ideals of R and the set $\mathcal{I}(\mathbb{M}_n(R))$ of all two sided ideals the ring $\mathbb{M}_n(R)$ of all $n \times n$ matrices with entries in R. [This gives another proof of the fact that if k is a division ring, then $\mathbb{M}_n(k)$ is a simple ring, as we have seen in the previous Exercise 1.1.11.]

The *center* of a ring R is

$$Z(R) := \{r \in R \mid rs = sr \text{ for every } s \in R\}.$$

1.2 Categories

It is easily verified that $Z(R)$ is a subring of R. When R is a division ring, the center $Z(R)$ of R is a field. Thus every division ring is a vector space over its center.

The first example of a division ring that is not a field is given by the division ring of *quaternions*, discovered by Hamilton in 1843. It is a 4-dimensional vector space $\mathbb{H}$ over the real field $\mathbb{R}$ with basis $\{1, i, j, k\}$, so that

$$\mathbb{H} = \mathbb{R} \oplus \mathbb{R}i \oplus \mathbb{R}j \oplus \mathbb{R}k.$$

Multiplication in $\mathbb{H}$ is given by

$$\begin{aligned} ij &= k, & jk &= i, & ki &= j, \\ ji &= -k, & kj &= -i, & ik &= -j, \\ i^2 &= -1, & j^2 &= -1, & k^2 &= -1 \end{aligned}$$

and then extending to a multiplication $\cdot : \mathbb{H} \times \mathbb{H} \to \mathbb{H}$ by $\mathbb{R}$-bilinearity. To show that every non-zero element $\alpha + \beta i + \gamma j + \delta k \in \mathbb{H}$ is invertible in $\mathbb{H}$, compute the element $\frac{1}{\alpha+\beta i+\gamma j+\delta k}$ multiplying both the numerator and the denominator by $\alpha - \beta i - \gamma j - \delta k$, and observe that $(\alpha + \beta i + \gamma j + \delta k)(\alpha - \beta i - \gamma j - \delta k) = \alpha^2 + \beta^2 + \gamma^2 + \delta^2$. The center of $\mathbb{H}$ is isomorphic to $\mathbb{R}$.

After Hamilton, Ferdinand Georg Frobenius [11] proved in 1877 that every finite-dimensional associative division $\mathbb{R}$-algebra D is isomorphic to $\mathbb{R}$, $\mathbb{C}$ or $\mathbb{H}$. (The exact definition of algebra over a field will be given in Definitions 2.2.7 and 2.2.8. Here we simply mean that if D is a division ring such that $Z(D) \supseteq \mathbb{R}$ up to isomorphism, and if the dimension of D as a vector space over $\mathbb{R}$ is finite, then D is isomorphic to $\mathbb{R}$, $\mathbb{C}$ or $\mathbb{H}$.) The first example of a non-commutative division ring infinite-dimensional over its center was given by Hilbert.

Wedderburn [22] proved in 1905 that every finite division ring is a field. Using this result, show that:

Exercise 1.1.13
Every finite integral domain is a field.

1.2 Categories

Definition 1.2.1

A *category* $\mathcal{C}$ consists of:

(i) a class $\mathrm{Ob}\,\mathcal{C}$, whose elements will be called the *objects* of $\mathcal{C}$;
(ii) for each pair (A, B) of objects of $\mathcal{C}$, a set $\mathrm{Hom}_\mathcal{C}(A, B)$, whose elements will be called *morphisms* of A into B;

(iii) for each triple (A, B, C) of objects of $\mathcal{C}$, a mapping

$$\circ \colon \mathrm{Hom}_{\mathcal{C}}(B, C) \times \mathrm{Hom}_{\mathcal{C}}(A, B) \to \mathrm{Hom}_{\mathcal{C}}(A, C),$$

called *composition*.

Before stating the axioms for categories, we introduce some notation. Instead of writing $f \in \mathrm{Hom}_{\mathcal{C}}(A, B)$, we will often write $f \colon A \to B$. For the composition $\circ \colon \mathrm{Hom}_{\mathcal{C}}(B, C) \times \mathrm{Hom}_{\mathcal{C}}(A, B) \to \mathrm{Hom}_{\mathcal{C}}(A, C)$ and morphisms $f \colon A \to B$, $g \colon B \to C$, we will denote the composite morphism by gf or $g \circ f$.
The axioms to which a category must satisfy are:

(1) If A, B, C, D are objects of $\mathcal{C}$ and $f \colon A \to B, g \colon B \to C, h \colon C \to D$ are morphisms, then $(hg)f = h(gf)$ (associativity of composition).
(2) For every $A \in \mathrm{Ob}\,\mathcal{C}$, there exists an element of $\mathrm{Hom}_{\mathcal{C}}(A, A)$, which we will denote 1_A, such that $f \circ 1_A = f$ and $1_A \circ g = g$ for every $B \in \mathrm{Ob}\,\mathcal{C}$, $f \in \mathrm{Hom}_{\mathcal{C}}(A, B)$ and $g \in \mathrm{Hom}_{\mathcal{C}}(B, A)$.

Categories were introduced by S. Eilenberg and S. MacLane in 1945 [8].

Examples 1.2.2
(1) *The category* **Set**. The objects of **Set** are all sets. If A, B are sets, the morphisms $f \colon A \to B$ are exactly all mappings $f \colon A \to B$, that is $\mathrm{Hom}_{\mathbf{Set}}(A, B) := B^A$. The composition is the composition of mappings. Then **Set** is a category, in which $1_A \colon A \to A$ is the identity mapping defined by $1_A(a) = a$ for every $a \in A$.
(2) *The category* **Grp**. The objects of **Grp** are all groups. If G, H are groups, the morphisms $f \colon G \to H$ are the "usual" group morphisms of G into H, that is, the mappings $f \colon G \to H$ such that $f(xy) = f(x)f(y)$ for every $x, y \in G$. The composition is the composition of mappings. Then **Grp** turns out to be a category.
(3) *The category* **Rng**. The objects of **Rng** are all rings with identity. The morphisms $f \colon R \to S$ are the ring morphisms of R into S. The composition is the composition of mappings.
(4) *The category* k-**Vect**, where k is a field. The objects of k-**Vect** are all vector spaces over k. The morphisms $f \colon V \to W$ are the k-linear transformations of V into W. The composition is the composition of mappings.
(5) *The category* **Top**. The objects of **Top** are all topological spaces. The morphisms $f \colon X \to Y$ between two topological spaces X and Y are the continuous mappings of X into Y. The composition is the composition of mappings.
(6) *The dual category (or opposite category)*. Let $\mathcal{C}$ be a category. Let $\mathcal{C}^{\mathrm{op}}$ be the category with $\mathrm{Ob}(\mathcal{C}^{\mathrm{op}}) = \mathrm{Ob}(\mathcal{C})$, $\mathrm{Hom}_{\mathcal{C}^{\mathrm{op}}}(A, B) = \mathrm{Hom}_{\mathcal{C}}(B, A)$ for every $A, B \in \mathrm{Ob}(\mathcal{C}^{\mathrm{op}}) = \mathrm{Ob}(\mathcal{C})$, and, if $*$ denotes the composition in $\mathcal{C}$ and $\circ$ denotes the composition in $\mathcal{C}^{\mathrm{op}}$, $f * g = g \circ f$. Then $\mathcal{C}^{\mathrm{op}}$ turns out to be a category, called the *dual category*, or the *opposite category*, of $\mathcal{C}$.
(7) Let $(P, \leq)$ be a *preordered set*, that is, a set P with a reflexive and transitive relation $\leq$. Let $\mathcal{P}$ be the category with:

1.2 Categories

(i) $\operatorname{Ob}\mathcal{P} = P$;
(ii) for every $a, b \in P$, $\operatorname{Hom}_\mathcal{P}(a,b) = \begin{cases} \{(a,b)\} & \text{if } a \leq b \\ \emptyset & \text{if } a \not\leq b; \end{cases}$
(Here $\{(a,b)\}$ denotes the set whose unique element is the pair (a,b).)
(iii) for every $a, b, c \in P$, the composition

$$\circ\colon \operatorname{Hom}_\mathcal{P}(b,c) \times \operatorname{Hom}_\mathcal{P}(a,b) \to \operatorname{Hom}_\mathcal{P}(a,c)$$

is the unique mapping of the set $\operatorname{Hom}_\mathcal{P}(b,c) \times \operatorname{Hom}_\mathcal{P}(a,b)$ into the set $\operatorname{Hom}_\mathcal{P}(a,c)$.

[Let us stop for a while to consider how many mappings there are from a set A into a set B when A and B are finite sets, that is, what is the cardinality of the set $B^A := \{\, f \mid f \text{ is a mapping of } A \text{ into } B\,\}$. It is well known that if $|X|$ denotes the cardinality of a set X, then $|B^A| = |B|^{|A|}$.

In particular, for $A \neq \emptyset$ and $B = \emptyset$, then $|B^A| = |\emptyset|^{|A|} = 0$. To explain this, notice that a mapping $f\colon A \to B$ is a correspondence of A into B, that is, a subset f of $A \times B$, such that for every $a \in A$ there is a unique $b \in B$ with $(a,b) \in f$. Hence a mapping $f\colon A \to B$ is a subset $f \subseteq A \times B$, such that

$$a \in A \implies \exists b \in B \text{ with } (a,b) \in f \tag{1.1}$$

and

$$a \in A, b \in B, b' \in B, (a,b) \in f, (a,b') \in f \implies b = b'. \tag{1.2}$$

When $A \neq \emptyset$ and $B = \emptyset$, the unique subset of $A \times B = A \times \emptyset = \emptyset$ is $f = \emptyset$. There are elements $a \in A$, but the thesis in sentence (1.1) is always false, so that sentence (1.1) is always false. Thus $f = \emptyset$ is not a mapping of A into B. Hence there are no subsets $f \subseteq A \times B$ that are mappings for $A \neq \emptyset$ and $B = \emptyset$, i.e., $B^A = \emptyset$.

Similarly, for $A = \emptyset$ and B an arbitrary set, $|B^A| = |B|^0 = 1$. To see this, let us go back to the previous definition of a mapping $f\colon A \to B$: it is a subset $f \subseteq A \times B$ satisfying (1.1) and (1.2). When $A = \emptyset$, the unique subset of $A \times B = \emptyset \times B = \emptyset$ is $f = \emptyset$, the hypotheses in (1.1) and (1.2) are always false, so that both sentences are true. In other words, $f = \emptyset$ is the unique mapping of $A = \emptyset$ into B. The mapping f is called the *empty mapping*. Thus $B^\emptyset = \{\emptyset\}$ has cardinality one.]

Let us apply this remark to consider mappings

$$\operatorname{Hom}_\mathcal{P}(b,c) \times \operatorname{Hom}_\mathcal{P}(a,b) \to \operatorname{Hom}_\mathcal{P}(a,c)$$

for a, b, c in our preordered set P.

If $a \not\leq b$ or $b \not\leq c$, then either $\operatorname{Hom}_\mathcal{P}(b,c) = \emptyset$ or $\operatorname{Hom}_\mathcal{P}(a,b) = \emptyset$, so that in both cases $\operatorname{Hom}_\mathcal{P}(b,c) \times \operatorname{Hom}_\mathcal{P}(a,b) = \emptyset$. Thus there is a unique mapping $\operatorname{Hom}_\mathcal{P}(b,c) \times \operatorname{Hom}_\mathcal{P}(a,b) \to \operatorname{Hom}_\mathcal{P}(a,c)$: the empty mapping.

Hence the remaining case is $a \leq b$ and $b \leq c$. In this case $a \leq c$ by transitivity, so that $\operatorname{Hom}_\mathcal{P}(b,c)$, $\operatorname{Hom}_\mathcal{P}(a,b)$ and $\operatorname{Hom}_\mathcal{P}(a,c)$ are three sets all of cardinality one, so that again there is a unique mapping $\operatorname{Hom}_\mathcal{P}(b,c) \times \operatorname{Hom}_\mathcal{P}(a,b) \to \operatorname{Hom}_\mathcal{P}(a,c)$.

It is now easy to see that P with these sets of morphisms is a category. Notice that the identity $1_a\colon a \to a$ is the unique element (a,a) of $\operatorname{Hom}_\mathcal{P}(a,a)$ ($a \leq a$ by reflexivity.)

(8) The *category* **Set**$_*$ *of pointed sets* has as objects all pairs (X, x_0), where X is a set and x_0 is an element of X. Morphisms of (X, x_0) into (Y, y_0) are the mappings $f\colon X \to Y$ such that $f(x_0) = y_0$.

(9) The category **Ab** of abelian groups has as objects all abelian additive groups, as morphisms $G \to H$ the group morphisms $G \to H$, and as composition the composition of mappings.

A category $\mathcal{C}$ is *small* if its class $\text{Ob}(\mathcal{C})$ of objects is a set.

Exercise 1.2.3

In Example (7), we have seen that if $(P, \leq)$ is a preordered set, it is possible to construct a category $\mathcal{P}$ with $\text{Ob}(\mathcal{P}) = P$ and at most one morphism between any two objects of $\mathcal{P}$ (there is a morphism $a \to b$ between any two objects of $\mathcal{P}$ if and only if $a \leq b$). The category $\mathcal{P}$ is small.

(a) Let $\mathcal{C}$ be a small category with the property that, for every $C, C' \in \text{Ob}(\mathcal{C})$, the set $\text{Hom}_\mathcal{C}(C, C')$ has cardinality ≤ 1. Define a relation ρ on the set $\text{Ob}(\mathcal{C})$ by setting, for every $C, C' \in \text{Ob}(\mathcal{C})$, $C\rho C'$ if the set $\text{Hom}_\mathcal{C}(C, C')$ is not empty. Show that $(\text{Ob}(\mathcal{C}), \rho)$ is a preordered set.

Hence preordered sets coincide, essentially, with small categories with at most one morphism between any two objects.

(b) Let $\mathcal{C}$ be any category and P any subset of $\text{Ob}(\mathcal{C})$. Define a relation ρ on P by setting, for every $C, C' \in P$, $C\rho C'$ if the set $\text{Hom}_\mathcal{C}(C, C')$ is not empty. Show that (P, ρ) is a preordered set.

Exercise 1.2.4

Now that we have seen that the concept of preorder on a set is a natural concept, we will describe preorders.

The main examples of preorders are equivalence relations (= symmetric preorders) and partial orders (= antisymmetric preorders). In this exercise, we will show that, conversely, every preorder can be constructed from an equivalence relation and a partial order.

Let X be a set.

(a) Let ρ be a preorder on X. Define a relation $\sim_\rho$ on X by setting, for every $x, y \in X$, $x \sim_\rho y$ if $x\rho y$ and $y\rho x$. Show that $\sim_\rho$ is an equivalence relation on X.

(b) Define a relation $\leq_\rho$ on the quotient set $X/\sim_\rho = \{[x]_{\sim_\rho} \mid x \in X\}$, by setting, for every $x, y \in X$, $[x]_{\sim_\rho} \leq_\rho [y]_{\sim_\rho}$ if $x\rho y$. Show that the relation $\leq_\rho$ is well defined and is a partial order on $X/\sim_\rho$.

(c) Show that there is a one-to-one correspondence between the set of all preorders ρ on X and the set of all pairs $(\sim, \leq)$, where $\sim$ is an equivalence relation on X and $\leq$ is a partial order on the quotient set $X/\sim$. It associates to every preorder ρ on X the pair $(\sim_\rho, \leq_\rho)$. Conversely, for any such pair $(\sim, \leq)$, the corresponding preorder $\rho_{(\sim,\leq)}$ on X is defined, for every $x, y \in X$, by $x\rho_{(\sim,\leq)} y$ if $[x]_\sim \leq [y]_\sim$.

Thus, a preorder ρ on X can be essentially seen as a partition $X/\sim_\rho$ of X, such that any two elements in the same block of the partition are ρ one to each other, and the partition $X/\sim_\rho$ is partially ordered by the partial order on $X/\sim_\rho$ induced by the preorder ρ.

1.3 Functors

Definition 1.3.1

Let $\mathcal{C}$ and $\mathcal{D}$ be categories. A *functor* (or a *covariant functor*) $F: \mathcal{C} \to \mathcal{D}$ assigns to every object $C \in \text{Ob}\,\mathcal{C}$ an object $F(C) \in \text{Ob}\,\mathcal{D}$, and to every morphism $f: C \to C'$ in $\mathcal{C}$ a morphism $F(f): F(C) \to F(C')$ in $\mathcal{D}$, and the following axioms are satisfied:

1.3 Functors

(i) if $f: C \to C'$ and $g: C' \to C''$ are morphisms in $\mathcal{C}$, then

$$F(g \circ f) = F(g) \circ F(f);$$

(ii) $F(1_C) = 1_{F(C)}$ for every $C \in \mathrm{Ob}\,\mathcal{C}$.

Definition 1.3.2

Let $\mathcal{C}$ and $\mathcal{D}$ be categories. A *contravariant functor* $F: \mathcal{C} \to \mathcal{D}$ assigns to every $C \in \mathrm{Ob}\,\mathcal{C}$ an object $F(C) \in \mathrm{Ob}\,\mathcal{D}$, and to every morphism $f: C \to C'$ a morphism $F(f): F(C') \to F(C)$, and the following axioms are satisfied: (i) $F(g \circ f) = F(f) \circ F(g)$ for every $f: C \to C'$ and $g: C' \to C''$; (ii) $F(1_C) = 1_{F(C)}$ for every $C \in \mathrm{Ob}\,\mathcal{C}$.

Examples 1.3.3
(1) Clearly, a contravariant functor $F: \mathcal{C} \to \mathcal{D}$ is essentially the same thing as a covariant functor $F: \mathcal{C}^{\mathrm{op}} \to \mathcal{D}$, where $\mathcal{C}^{\mathrm{op}}$ denotes the dual category.
(2) *The identity functor.* Let $\mathcal{C}$ be an arbitrary category. The functor $I: \mathcal{C} \to \mathcal{C}$, defined by $I(C) = C$ for every object $C \in \mathrm{Ob}\,\mathcal{C}$ and $I(f) = f$ for every morphism $f: A \to B$ in $\mathcal{C}$, is trivially a covariant functor. It is also denoted as $I_\mathcal{C}$ or $1_\mathcal{C}$.
(3) *The opposite functor.* The functor $F: \mathcal{C} \to \mathcal{C}^{\mathrm{op}}$, defined by $F(C) = C$ for every object $C \in \mathrm{Ob}\,\mathcal{C}$ and $F(f) = f$ for every morphism $f: A \to B$ in $\mathcal{C}$, is a contravariant functor.
(4) Let **Grp** be the category of groups (Example 1.2.2(2)). Fix a group $\overline{G} \in \mathrm{Ob}\,\mathbf{Grp}$, and define $F: \mathbf{Grp} \to \mathbf{Set}$ as follows: on objects,

$$F(G) = \mathrm{Hom}_{\mathbf{Grp}}(\overline{G}, G)$$

for every $G \in \mathrm{Ob}\,\mathbf{Grp}$; on morphisms,

$$F(f): \mathrm{Hom}_{\mathbf{Grp}}(\overline{G}, G) \to \mathrm{Hom}_{\mathbf{Grp}}(\overline{G}, G')$$

is, for every morphism $f: G \to G'$, the mapping $F(f): \alpha \mapsto f \circ \alpha$. Then F is a functor of **Grp** into **Set**, because if $f: G \to G'$, $g: G' \to G''$ are morphisms in **Grp**, then for every $\alpha \in \mathrm{Hom}_{\mathbf{Grp}}(\overline{G}, G)$,

$$F(g \circ f)(\alpha) = (g \circ f) \circ \alpha = g \circ (f \circ \alpha) = (F(g) \circ F(f))(\alpha),$$

so that $F(g \circ f) = F(g) \circ F(f)$. Moreover, $F(1_G)(\alpha) = 1_G \circ \alpha = \alpha$ for all $\alpha \in \mathrm{Hom}_{\mathbf{Grp}}(\overline{G}, G)$, so that $F(1_G) = 1_{\mathrm{Hom}_{\mathbf{Grp}}(\overline{G}, G)}$.
(5) The forgetful functor. Let $F: \mathbf{Top} \to \mathbf{Set}$ be the functor that assigns to every topological space $(X, \tau_X) \in \mathrm{Ob}(\mathbf{Top})$ the set $X \in \mathrm{Ob}(\mathbf{Set})$, and to every continuous mapping $f: (X, \tau_X) \to (Y, \tau_Y)$ the mapping $f: X \to Y$. Then F is a coviariant functor, called the forgetful functor, because it *forgets* the topological structure of a topological space (X, τ_X).
(6) Let $C: \mathbf{Top} \to \mathbf{Rng}$ be the functor that assigns to every topological space $X \in \mathrm{Ob}(\mathbf{Top})$ the ring

$$C(X, \mathbb{R}) = \{\varphi: X \to \mathbb{R} \mid \varphi \text{ is continuous}\}$$

of all continuous functions of the topological space X into the topological space $\mathbb{R}$ of the real numbers with the usual topology. Notice that $C(X, \mathbb{R})$ is a commutative ring with respect to the operations defined by $(\varphi + \psi)(x) = \varphi(x) + \psi(x)$ and $(\varphi\psi)(x) = \varphi(x)\psi(x)$ for every

$\varphi, \psi \in C(X, \mathbb{R})$ and every $x \in X$. Then C is a contravariant functor that assigns to every morphism f in **Top**, that is, to every continuous mapping $f \colon X \to Y$, the ring morphism $C(f) \colon C(Y, \mathbb{R}) \mapsto C(X, \mathbb{R})$, defined by $C(f)(\varphi) = \varphi \circ f$ for every $\varphi \in C(Y, \mathbb{R})$.
(7) Let k be a field. Let k-**Vect** be the category of all vector spaces over k (Example 1.2.2 (4)). Consider the functor $D \colon k$-**Vect** $\to k$-**Vect** that assigns to every $V \in \mathrm{Ob}(k$-**Vect**$)$ its dual space $D(V) = V^* = \mathrm{Hom}_k(V, k)$ and to every linear transformation $f \colon V \to W$ the transpose $D(f) = f^* \colon W^* \to V^*$, defined by $f^*(\varphi) = \varphi \circ f$ for all $\varphi \in W^*$. The functor D is contravariant.
(8) *Composition of functors.* Let $\mathcal{C}, \mathcal{D}, \mathcal{E}$ be categories and $F \colon \mathcal{C} \to \mathcal{D}, G \colon \mathcal{D} \to \mathcal{E}$ be functors. The composite functor $GF \colon \mathcal{C} \to \mathcal{E}$ is defined in the obvious way: $GF(C) = G(F(C))$ for every object C in $\mathcal{C}$; $GF(f) = G(F(f))$ for every morphism $f \colon C \to C'$ in $\mathcal{C}$. Notice that if F and G are both covariant or both contravariant, the composite functor GF is covariant. If one is covariant and the other is contravariant, the composite functor GF is contravariant.
(9) *Isomorphism of categories.* An *isomorphism* of a category $\mathcal{C}$ into a category $\mathcal{D}$ is a functor $F \colon \mathcal{C} \to \mathcal{D}$ such that there exists a functor $G \colon \mathcal{D} \to \mathcal{C}$ with GF the identity functor $I_\mathcal{C} \colon \mathcal{C} \to \mathcal{C}$ and FG the identity functor $I_\mathcal{D} \colon \mathcal{D} \to \mathcal{D}$. If there is an isomorphism $F \colon \mathcal{C} \to \mathcal{D}$ we will say that the two categories $\mathcal{C}$ and $\mathcal{D}$ are *isomorphic*.

1.4 Modules

Let R be a fixed ring with identity 1_R. It is possible to define left modules over the ring R in two equivalent ways. Recall that, for every abelian group M, we denote by $\mathrm{End}(M)$ the endomorphism ring of M.

Definition 1.4.1

A *left R-module* (or *left module over the ring R*) is a triple $(M, +, \cdot)$, where $(M, +)$ is an additive abelian group and $\cdot \colon R \times M \to M, \cdot \colon (r, m) \mapsto rm$, is a mapping, called *left scalar multiplication*, with the following properties for every $r, r' \in R$, and every $m, m' \in M$:

(i) $r(r'm) = (rr')m$;
(ii) $(r + r')m = rm + r'm$;
(iii) $r(m + m') = rm + rm'$;
(iv) $1_R m = m$.

Definition 1.4.2

A *left R-module* (or *left module over the ring R*) is a triple $(M, +, \lambda)$, where $(M, +)$ is an additive abelian group and $\lambda \colon R \to \mathrm{End}(M)$ is a ring homomorphism of R into the ring $\mathrm{End}(M)$ of all endomorphisms of the abelian group $(M, +)$.

These two definitions are equivalent in the following sense. Assume that $(M, +, \cdot)$ is a module defined as in Definition 1.4.1. Let $\lambda \colon R \to \mathrm{End}(M)$ be the mapping defined by $\lambda(r)(m) = rm$ for every $r \in R, m \in M$. Then λ is a ring morphism of R into the ring $\mathrm{End}(M)$ of all endomorphisms of the abelian group

1.4 Modules

$(M, +)$. To see it, we must check four conditions: that $\lambda(r) \in \text{End}(M)$ for every $r \in R$, $\lambda(r + r') = \lambda(r) + \lambda(r')$, $\lambda(rr') = \lambda(r)\lambda(r')$, $\lambda(1_R) = 1_{\text{End}(M)}$. These four conditions follow from properties (iii), (ii), (i), (iv) of Definition 1.4.1 respectively. Thus $(M, +, \lambda)$ becomes a left module as defined in Definition 1.4.2.

Conversely, let $(M, +, \lambda)$ be a module as in Definition 1.4.2. Define a scalar multiplication $\cdot : R \times M \to M$ setting $\cdot : (r, m) \mapsto rm := \lambda(r)(m)$ for every $r \in R$, $m \in M$. Then from the fact that λ maps R into $\text{End}(M)$ and respects addition, multiplication and the identity, we get the four properties (iii), (ii), (i), (iv) of Definition 1.4.1, that is, $(M, +, \cdot)$ is a left module in the sense of Definition 1.4.1.

Thus the two definitions of a left module are logically equivalent, and we will use both, depending on the convenience.

Definition 1.4.3

Let R be a ring and let M, N be left R-modules. A *module morphism* (or *module homomorphism*) of M into N is a mapping $f : M \to N$ such that, for every $x, y \in M$ and every $r \in R$, $f(x + y) = f(x) + f(y)$ and $f(rx) = rf(x)$.

We can be very precise and describe the logical equivalence of the two Definitions 1.4.1 and 1.4.2 of left R-modules in categorical terms. Define a category R-Mod$_1$ in which: the objects are all modules $(M, +, \cdot)$ defined as in Definition 1.4.1; the morphisms $f : (M, +, \cdot) \to (M', +, \cdot)$ in R-Mod$_1$ are the module morphisms as defined in Definition 1.4.3; the composition in R-Mod$_1$ is the composition of mappings. Similarly, we can define another category R-Mod$_2$ whose objects are all modules $(M, +, \lambda)$ defined as in Definition 1.4.2. A morphism $f : (M, +, \lambda_M) \to (M', +, \lambda'_{M'})$ in R-Mod$_2$ is a group morphism $f : (M, +) \to (M', +)$ such that the diagram

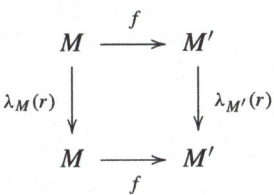

is commutative for every $r \in R$, that is, such that $f \circ \lambda_M(r) = \lambda_{M'}(r) \circ f$ for every $r \in R$. The composition in R-Mod$_2$ is the composition of mappings. Then the position $(M, +, \cdot) \mapsto (M, +, \lambda)$ can be extended to a functor $F : R$-Mod$_1 \to R$-Mod$_2$, and the position $(M, +, \lambda) \mapsto (M, +, \cdot)$ can be extended to a functor $G : R$-Mod$_2 \to R$-Mod$_1$. These two functors F and G are one the inverse of the other, so that the categories R-Mod$_1$ and R-Mod$_2$ are isomorphic (Example 1.3.3(9)).

When R is a division ring D, left D-modules are usually called *left vector spaces* over the division ring D.

Let us pass to define *right* modules. The definition is similar to that of left modules, but the scalars acts on the right instead of on the left.

Definition 1.4.4

A *right R-module* (or *right module over the ring R*) is a triple $(M, +, \cdot)$, where $(M, +)$ is an additive abelian group and $\cdot \colon M \times R \to M$, $\cdot \colon (m, r) \mapsto mr$, is a mapping, called *right scalar multiplication*, with the following properties for every $r, r' \in R$, and every $m, m' \in M$:

(i) $(mr)r' = m(rr')$;
(ii) $m(r + r') = mr + mr'$;
(iii) $(m + m')r = mr + m'r$;
(iv) $m1_R = m$.

For another definition, analogue to that of Definition 1.4.2, we need the notion of ring anti-homomorphism.

Definition 1.4.5

Let R and S be rings. A *ring anti-homomorphism* $f \colon R \to S$ is a mapping of the set R into the set S such that:

(i) $f(r + r') = f(r) + f(r')$ for every $r, r' \in R$;
(ii) $f(rr') = f(r')f(r)$ for every $r, r' \in R$;
(iii) $f(1_R) = 1_S$.

Example 1.4.6
Let k be a field, n be a positive integer, and $\mathbb{M}_n(k)$ be the ring of $n \times n$ matrices with entries in k. The *transposition* $t \colon \mathbb{M}_n(k) \to \mathbb{M}_n(k)$ defined by $A \mapsto A^t$ (where A^t is the transpose of A) is a ring *anti-isomorphism*, that is, a ring anti-homomorphism that is also a bijective mapping.

Example 1.4.7
If $(R, +, \cdot)$ is a ring, its *opposite ring* is the ring $(R, +, \circ)$, where $\circ \colon R \times R \to R$ is the new operation on the set R defined by $r \circ r' = r' \cdot r$. Usually, if R is a ring, its opposite ring is denoted by R^{op}. It is easlily see that R^{op} is really a ring for every ring R. Then the identity mapping $\iota_R \colon R \to R$, defined by $r \in R \mapsto r$, viewed as a mapping $R \to R^{op}$, is an anti-isomorphism of R onto R^{op}.

Definition 1.4.8

Let R be a ring. A *right R-module* $(M, +, \rho)$ is an abelian group $(M, +)$ together with a ring anti-homomorphism $\rho \colon R \to \text{End}(M)$.

Also for right modules, it is easy to see that the two Definitions 1.4.4 and 1.4.8 give the same structures, or, if we want to be more precise, that the two corresponding categories are isomorphic. Both for right modules and for left modules

1.4 Modules

we will not distinguish between the two possible definitions. We will consider the category R-Mod of all left R-modules and we will use both the definition with the left scalar multiplication or with the ring morphism $R \to \mathrm{End}(M)$. Similarly, on the other side, we will consider the category Mod-R of all right R-modules and we will use both the definition with the right scalar multiplication or with the ring anti-homomorphism $R \to \mathrm{End}(M)$, as it is more convenient.

▶ **Remark 1.4.9** It is clear that ring anti-homomorphisms

$$R \to \mathrm{End}(M)$$

and ring homomorphisms

$$R^{\mathrm{op}} \to \mathrm{End}(M)$$

coincide. It follows that right R-modules and left R^{op}-modules are exactly the same thing. Similarly, left R-modules coincide with right R^{op}-modules. Also, notice that if the ring R is commutative, a mapping

$$R \to \mathrm{End}(M)$$

is a ring homomorphism if and only if it is a ring anti-homomorphism. It follows that right modules and left modules coincide over a commutative ring R. If we want to be more precise, we can use the categorical language, and say that there is an isomorphism of categories between the category of all right R-modules and the category of all left R^{op}-modules. Similarly, for R commutative, the category of all right R-modules and the category of all left R-modules are isomorphic, that is, a right R-module structure on an abelian group M is the same as a left R-module structure on M.

If M is a right R-module, we will usually denote it by M_R, and if M is a left R-module, we will denote it by $_RM$. That is, we will write the ring R of "scalars" on the side on which it acts.

If $f \colon M_R \to N_R$ is a module morphism, we will say that f is a *monomorphism* if f is injective, an *epimorphism* if it is surjective, an *isomorphism* if it is a bijection, an *endomorphism* of M_R if $M_R = N_R$, an *automorphism* of M_R if it is an endomorphism of M_R and a bijection.

Example 1.4.10
Let $(G, +)$ be an abelian group and $\mathrm{End}(G)$ its endomorphism ring (page 5). We have already seen on page 3 that there is a unique ring homomorphism $\lambda \colon \mathbb{Z} \longrightarrow \mathrm{End}(G)$. Equivalently, there is a unique left $\mathbb{Z}$-module structure on any abelian group G. The scalar multiplication $\cdot \colon \mathbb{Z} \times M \to M$

of M is given by $nx =$ "n-th multiple of x in the additive group M" for every $n \in \mathbb{Z}$, $x \in M$. That is,

$$nx = \begin{cases} \underbrace{x + \cdots + x}_{n \text{ times}} & \text{if } n > 0 \\ 0_M & \text{if } n = 0 \\ \underbrace{(-x) + (-x) + \cdots + (-x)}_{-n \text{ times}} & \text{if } n < 0 \end{cases}$$

Thus, left $\mathbb{Z}$-modules and abelian groups coincide. If we want to be more precise, it is easily seen that the category **Ab** of all abelian groups (Example 1.2.2(9)) is isomorphic to the category $\mathbb{Z}$-Mod of all left $\mathbb{Z}$-modules, an isomorphism $F\colon \mathbb{Z}$-Mod $\to$ **Ab** being the forgetful functor F.

Example 1.4.11
Let M_R and N_R be right R-modules. Let

$$\text{Hom}(M_R, N_R)$$

be the set of all module morphisms $f\colon M_R \to N_R$. Then $\text{Hom}(M_R, N_R)$ is an abelian group with respect to the addition defined by $(f + g)(x) = f(x) + g(x)$ for every $f, g \in \text{Hom}(M_R, N_R)$ and $x \in M_R$.

Fix a module M_R. It is possible to define a covariant functor

$$\text{Hom}(M_R, -)\colon \text{Mod-}R \to \textbf{Ab}.$$

It is sufficient to associate to every right R-module N_R the abelian group $\text{Hom}(M_R, N_R)$, and to every R-module morphism $f\colon N_R \to N'_R$ the group morphism

$$\text{Hom}(M_R, f)\colon \text{Hom}(M_R, N_R) \to \text{Hom}(M_R, N'_R),$$

defined by $\text{Hom}(M_R, f)(\alpha) = f \circ \alpha$ (composition of mappings) for every $\alpha \in \text{Hom}(M_R, N_R)$.

Dually, fix a module N_R. There is a *contravariant* functor

$$\text{Hom}(-, N_R)\colon \text{Mod-}R \to \textbf{Ab}.$$

It associates to every right R-module M_R the abelian group

$$\text{Hom}(M_R, N_R),$$

and to every R-module morphism $f\colon M_R \to M'_R$ the group morphism

$$\text{Hom}(f, N_R)\colon \text{Hom}(M'_R, N_R) \to \text{Hom}(M_R, N_R)$$

defined by $\text{Hom}(f, N_R)(\beta) = \beta \circ f$ for every $\beta \in \text{Hom}(M'_R, N_R)$.

Example 1.4.12
Let R be a ring and let M_R be a right R-module. The abelian group $\text{End}(M_R) := \text{Hom}(M_R, M_R)$ of the Example 1.4.11 becomes a ring with respect to the multiplication defined by

$$(fg)(x) = f(g(x)) \qquad \text{for all } f, g \in \text{End}(M_R),\ x \in M_R.$$

1.4 Modules

It is called the *endomorphism ring* of M_R. We can give M a left $\text{End}(M_R)$-module structure. For this, define the left scalar multiplication

$$*\colon \text{End}(M_R) \times M \to M, \quad *\colon (f, m) \mapsto f * m,$$

setting

$$f * x = f(x) \quad \text{for every } f \in \text{End}(M_R), x \in M.$$

It is easily seen that the axioms for left $\text{End}(M_R)$-modules hold. For example, if $f \in \text{End}(M_R)$, and $x, y \in M_R$, we have

$$f * (x + y) = f(x + y) = f(x) + f(y) = f * x + f * y,$$

because f in an endomorphism. In this case, the ring homomorphism $\lambda\colon \text{End}(M_R) \to \text{End}(M_\mathbb{Z})$ corresponding to the other definition of left $\text{End}(M_R)$-module (Definition 1.4.2) is the inclusion mapping.

In particular, if k is a field and V_k is a vector space over k of finite dimension n, then V_k turns out to be not only a right module over k, but also a left module over $\text{End}(V_k) \cong M_n(k)$, the ring of $n \times n$ matrices with entries in k.

Example 1.4.13
If R is a ring and $_R M$ is a left R-module, it is possible to give the left module $_R M$ a right $\text{End}(_R M)$-module structure. For left modules, it is usually more convenient to write mappings on the right. That is, if $_R M$ is a left R-module and $f\colon {_R M} \to {_R M}$ is an endomorphism, we will denote the image of an element $x \in {_R M}$ by $(x)f$. The right $\text{End}(_R M)$-module structure on $_R M$ is given defining, as right scalar multiplication $*\colon M \times \text{End}(_R M) \to M$, $*\colon (x, f) \mapsto x * f$,

$$x * f = (x)f \quad \text{for every } f \in \text{End}(M_R),\ x \in M.$$

Example 1.4.14
Let K be a division ring, and let D be a *sub-division ring* of K, that is, a subring D of K such that $d^{-1} \in D$ for every $d \in D$, $d \neq 0$. The multiplication on K is a mapping $\cdot\colon K \times K \to K$, which can be restricted to a scalar multiplication $D \times K \to K$. It is immediate to see that K, with this scalar multiplication, is a left D-module $_D K$ in a natural way. Similarly, but on the other side, the multiplication $\cdot\colon K \times K \to K$ on the division ring K can be restricted to a scalar multiplication $K \times D \to K$ and K, with this scalar multiplication, becomes a right D-module K_D. We will discuss later the possibility of talking of *dimension* $\dim(V_D)$ of a right vector space V_D or $\dim(_D V)$ of a left vector space $_D V$, but it is possible, though it is not easy, to construct, for any $1 < n, m \leq \infty$, a division ring K with a sub-division ring D for which $\dim(K_D) = n$ and $\dim(_D K) = m$ [25].

▶ **Remark 1.4.15** The definition of left R-modules, which correspond to ring homomorphisms, seems more natural than that of right R-modules, corresponding to the less natural notion of ring anti-homomorphism. The reason of this lies in the fact that we use to write mappings on the left, and not on the right. To be more precise, let A and B be sets and assume that we have a mapping $f\colon A \to B$. Then we use to denote the image of an element $a \in A$ by $f(a)$. Also, if $f\colon A \to B$ and $g\colon B \to C$ are two mappings, we denote their composite mapping by $g \circ f$, which is the mapping that sends an element $a \in A$ to $(g \circ f)(a) = g(f(a))$. The choice of this notation was arbitrary, and we could write mappings on the right. For a mapping $f: A \to B$, it is possible to denote the image of an element $a \in A$ by $(a)f$, with the mapping f on the right of the elements a on which f acts. In this case, if $f: A \to B$ and $g: B \to C$ are two mappings, it is more natural to denote the composite mapping by $f \circ g$, because it sends the element $a \in A$ to $((a)f)g$. Notice that, in some settings, mappings *are* denoted on the right. For instance, in group theory, it is common to denote an action g, for instance, conjugation, on an element a in the form a^g. Here g is written as an exponent, that is, on the right of the elements a on which it acts.

If, for any reason, we write mappings on the right, then right R-modules correspond to ring homomorphisms $R \to \text{End}(M)$, and left R-modules correspond to ring anti-homomorphisms of R into $\text{End}(M)$. If, in the ring $\text{End}(M)$ of all endomorphisms of a non-zero abelian group M, we write endomorphisms on the right, then the ring of all endomorphisms of M with endomorphisms written on the right is $\text{End}(M)^{op}$.

From now on, we will always write, as usual, mappings on the left, and most modules we will consider will be right modules M_R.

Exercise 1.4.16
Let M be a right R-module, $x, y \in M, r, s \in R$. Show that:
(i) $0_M r = 0_M$.
(ii) $x 0_R = 0_M$.
(iii) $(-x)r = -(xr)$ and $x(-r) = -(xr)$. We will denote the element $(-x)r = x(-r) = -(xr)$ by $-xr$.
(iv) $(x - y)r = xr - yr$ and $x(r - s) = xr - xs$.

[Recall that in any additive group G, one writes $a - b$ to denote the sum of a and the opposite of b. That is, $a - b := a + (-b)$.]

Notice that in this exercise we have used only properties (i), (ii) and (iii) of Definition 1.4.1. Hence, what we have proved also holds for "nonunitary modules", that is, the algebraic structures considered in the following Exercise 1.4.17.

Also notice that over the ring $R = 0$, all right R-modules are the trivial module, that is, the zero module with one element, because for every x in such a module M_R, $x = x 1_R = x 0_R = 0_M$.

Exercise 1.4.17
Let R be a ring with identity, $(M, +)$ an additive abelian group and $\cdot \colon R \times M \to M$, $\cdot \colon (r, m) \mapsto rm$, a mapping that satisfies properties (i), (ii) and (iii) of Definition 1.4.1. Let M_0 be the set of all $x \in M$ with $1_R x = 0_M$, and M_1 be the set of all $x \in M$ with $1_R x = x$. Show that:
(i) M_0 and M_1 are subgroups of M.
(ii) M is the direct sum of M_0 and M_1 as abelian groups.
(iii) $rx = 0_M$ for every $r \in R$ and $x \in M_0$.
(iv) M_1 is a left R-module with respect to the left scalar multiplication induced by $\cdot$.

Sometimes "modules" are defined as the algebraic structures satisfying properties (i), (ii) and (iii) of Definition 1.4.1, and those satisfying property (iv) also are called "unitary modules". Thus every "non-unitary module" is the direct sum of a "module" over which R acts trivially and a "unitary module".

1.5 Bimodules

Definition 1.5.1

Let R and S be rings. An *R-S-bimodule* is an ordered quadruple $(M, +, \cdot, *)$, where $(M, +, \cdot)$ is a left R-module $_RM$, $(M, +, *)$ is a right S-module M_S, and the two module structures are compatible in the sense that

$$(r \cdot x) * s = r \cdot (x * s) \qquad \text{for every } r \in R, s \in S, x \in M.$$

We will denote an R-S-bimodule by $_RM_S$.

Exercise 1.5.2
If R is a commutative ring, every right R-module M_R can be viewed as an R-R-bimodule. More precisely, for any (non-necessarily commutative) ring $(R, +, \circ)$, let $(M_R, +, *)$ be a right R-module. Consider $(M, +, \cdot)$, where $\cdot : R \times M \to M$ is the left scalar multiplication defined, for every $r \in R$ and $x \in M$, by $r \cdot x = x * r$.
(1) Show that, if R is commutative, then $(M, +, \cdot, *)$ is a bimodule $_RM_R$.
   Now assume R non-necessarily commutative. In this case:
(2) Is $(M, +, \cdot)$ a left module $_RM$?
(3) Is $(M, +, \cdot, *)$ a bimodule $_RM_R$?
(4) Is $(M, +, \cdot)$ a left module $_{R^{op}}M$?
(3) Is $(M, +, \cdot, *)$ an R^{op}-R-bimodule $_{R^{op}}M_R$?

Example 1.5.3
If R is a ring, the multiplication $\cdot : R \times R \to R$ in the ring R can be viewed both as a right scalar multiplication and as a left scalar multiplication. Thus R has the structure of right module R_R and of left R-module $_RR$. It is easily checked that R is a R-R-bimodule.

Example 1.5.4
Let R be a ring and M_R a right R-module. Thus M is an additive abelian group, hence, by Example 1.4.10, M is also a left $\mathbb{Z}$-module. As right multiplication by an element $r \in R$ is a group morphism $M \to M$, that is, a module morphism $_{\mathbb{Z}}M \to _{\mathbb{Z}}M$, we have that $(nx)r = n(xr)$ for every $n \in \mathbb{Z}, x \in M, r \in R$. Hence every right R-module M_R is a $\mathbb{Z}$-R-bimodule $_{\mathbb{Z}}M_R$. Similarly, left R-modules $_RN$ are R-$\mathbb{Z}$-bimodules $_RN_{\mathbb{Z}}$.

Example 1.5.5
Let R be a ring and $(M_R, +, *)$ a right R-module. In Example 1.4.12, we have seen that M_R has a left $\text{End}(M_R)$-module structure defined by $f \cdot x = f(x)$ for every $f \in \text{End}(M_R)$ and every $x \in M_R$. The equality

$$(f \cdot x) * r = f(x) * r = f(x * r) = f \cdot (x * r)$$

shows that M_R is a End(M_R)-R-bimodule. Similarly, every left R-module $_RM$ is an R- End$(_RM)$-bimodule $_RM_{\text{End}(_RM)}$.

Example 1.5.6
Let R be a ring and $\mathbb{M}_n(R)$ be the ring of all $n \times n$ matrices with entries in R. Similarly for $\mathbb{M}_m(R)$. Let $\mathbb{M}_{n \times m}(R)$ be the abelian group of all $n \times m$ matrices with entries in R. Then $_{\mathbb{M}_n(R)}\mathbb{M}_{n \times m}(R)_{\mathbb{M}_m(R)}$ is a $\mathbb{M}_n(R)$-$\mathbb{M}_m(R)$-bimodule. Scalar multiplications are row by column multiplication.

Example 1.5.7
Let R be a ring. We already know that if M_R, N_R are right R-modules, the set Hom(M_R, N_R) all module morphisms $f : M_R \to N_R$ is an abelian group with respect to the addition defined by $(f + g)(x) = f(x) + g(x)$ for every $f, g \in$ Hom(M_R, N_R) and every $x \in M_R$ (Example 1.4.11).

If the ring R is commutative, we can define an R-module structure on Hom(M_R, N_R) via the scalar multiplication $\cdot$ defined by

$$(r \cdot f)(x) = (f(x))r \quad \text{for every } r \in R, \ f \in \text{Hom}(M_R, N_R) \text{ and } x \in M_R.$$

(It is easily verified that Hom(M_R, N_R) is a left R-module. For instance, for $f \in$ Hom(M_R, N_R) and $r, s \in R$, one has

$$((rs) \cdot f)(x) = (rs)(f(x)) = r(s(f(x))) = r((s \cdot f)(x)) = (r \cdot (s \cdot f))(x).$$

In this case, the covariant functor Hom$(M_R, -)$ (for every fixed module M_R) and the contravariant functor Hom$(-, N_R)$ (for every fixed module N_R) turn out to be functors Mod-$R \to R$-Mod (not only Mod-$R \to$ **Ab**.))

Let us go back to the case of R arbitrary, not necessarily commutative. In this case, the additive group Hom(M_R, N_R) is a left module over the endomorphism ring End(N_R) of N_R (Example 1.4.12). It suffices to define, for every $\varphi \in$ End(N_R) and every $f \in$ Hom(M_R, N_R), $\varphi f = \varphi \circ f$ (composite mapping of the mappings $f : M_R \to N_R$ and $\varphi : N_R \to N_R$.)

Similarly, the additive group Hom(M_R, N_R) becomes a right module over the endomorphism ring End(M_R) of M_R if one defines $f \cdot \alpha = f \circ \alpha$ (composite mapping of $\alpha : M_R \to M_R$ and $f : M_R \to N_R$) for every $f \in$ Hom(M_R, N_R) and every $\alpha \in$ End(M_R). Thus we have a left module $_{\text{End}(N_R)}$ Hom(M_R, N_R) and a right module Hom$(M_R, N_R)_{\text{End}(M_R)}$. It is easy to check that

$$\text{Hom}(M_R, N_R)$$

is a End(N_R)-End(M_R)-bimodule

$$_{\text{End}(N_R)} \text{Hom}(M_R, N_R)_{\text{End}(M_R)}.$$

In this case also, we have a covariant functor

$$\text{Hom}(M_R, -): \text{Mod-}R \to \text{Mod-End}(M_R)$$

for every module M_R and the contravariant functor

$$\text{Hom}(-, N_R): \text{Mod-}R \to \text{End}(N_R)\text{-Mod}$$

for every module N_R.

1.5 Bimodules

To recall the correct side (right or left), recall that $\text{Hom}(M_R, -)$ always preserves the arrows: it is covariant and transforms right modules over R into right modules over $\text{End}(M_R)$. On the contrary, $\text{Hom}(-, N_R)$ always reverses the arrows: it is contravariant and transforms right modules over R into left modules over $\text{End}(N_R)$.

Example 1.5.8
Assume that R, S, T are rings and that ${}_S M_R, {}_T N_R$ are bimodules. Then $\text{Hom}(M_R, N_R)$ becomes a left T-module defining

$$(tf)(x) = tf(x)$$

for every $t \in T$, $f \in \text{Hom}(M_R, N_R)$ and $x \in M$. Similarly, $\text{Hom}(M_R, N_R)$ becomes a right S-module defining

$$(fs)(x) = f(sx) \text{ for every } f \in \text{Hom}(M_R, N_R), \ s \in S \text{ and } x \in M.$$

Moreover, $\text{Hom}(M_R, N_R)$ is a T-S-bimodule ${}_T \text{Hom}(M_R, N_R)_S$. Again, notice that $\text{Hom}(-, -)$ is contravariant on the first variable and covariant on the second. That is, $\text{Hom}(M_R, -)$ preserves the arrows (${}_T N_R$ left T-module implies ${}_T \text{Hom}(M_R, N_R)$ left T-module) and $\text{Hom}(-, N_R)$ reverses the arrows (${}_S M_R$ left S-module implies $\text{Hom}(M_R, N_R)_S$ right S-module). Moreover, $\text{Hom}({}_S M_R, -)$ is a covariant functor Mod-$R \to$ Mod-S, and $\text{Hom}(-, {}_T N_R)$ is a contravariant functor Mod-$R \to T$-Mod.

Everything can be done on the opposite side: If ${}_R M_S$ and ${}_R N_T$ are bimodules, then ${}_S(\text{Hom}({}_R M_S, {}_R N_T))_T$ has a canonical S-T-bimodule structure. (How are the operations defined in this case?)

Also for bimodules, like for right modules and left modules, it is possible to give various equivalent definitions. In fact, it is easy to prove the following proposition:

Proposition 1.5.9
Let R and S be rings, $(M, +)$ be an additive abelian group, $\cdot \colon R \times M \to M$ and $* \colon M \times S \to M$ be mappings. Define $\lambda(r)(x) = r \cdot x$ and $\rho(s)(x) = x * s$ for every $r \in R$, $s \in S$ and $x \in M$. The following conditions are equivalent:

(i) *The quadruple $(M, +, \cdot, *)$ is an R-S-bimodule.*
(ii) *$(M, +, *)$ is a right S-module and $\lambda \colon R \to \text{End}(M_S)$ is a ring homomorphism.*
(iii) *$(M, +, \cdot)$ is a left R-module and $\rho \colon S \to \text{End}({}_R M)$ is a ring anti-homomorphism (here, as we have decided to do on page 18, we write mappings belonging to the ring $\text{End}({}_R M)$ on the left; if we write mappings belonging to $\text{End}({}_R M)$ on the right, then ρ turns out to be a ring homomorphism).*

For example, assume (i) true. As M is an R-S-bimodule, we have in particular that $(M, +, *)$ is a right S-module. Since $(M, +, \cdot)$ is also a left R-module, we have the ring homomorphism $\lambda': R \longrightarrow \text{End}(M)$ defined by $\lambda'(r)(x) = r \cdot x$ for every $r \in R$ and $x \in M$. Here $\text{End}(M)$ is the endomorphism ring of M as an additive group. But the image of λ' is contained in $\text{End}(M_S)$, because $\lambda'(r)$ is a right S-module morphism (for every $s \in S$, we have that $\lambda'(r)(x*s) = r \cdot (x*s) = (r \cdot x) * s = (\lambda'(r)(x)) * s$). The ring morphism $\lambda: R \to \text{End}(M_S)$ defined in the statement is obtained by restricting the range of $\lambda': R \to \text{End}(M)$ to $\text{End}(M_S)$. Thus (ii) holds.

Conversely, if (ii) holds, then M is both a left R-module and a right S-module $(M, +, *)$ in which the left scalar multiplication $\cdot$ is defined by $r \cdot x = \lambda(r)(x)$ for every $r \in R$ and $x \in M$. Moreover, $(r \cdot x)*s = (\lambda(r)(x))*s = \lambda(r)(x*s) = r \cdot (x*s)$ for every $r \in R$, $x \in M$ and $s \in S$. Thus M is an R-S-bimodule.

1.6 Submodules and Quotients

Definition 1.6.1

Let R be a ring and M_R a right R-module. A subset N of M_R is said to be a *submodule* of M_R if N is non-empty and $x - y \in N$, $xr \in N$ for every $x, y \in N_R$, $r \in R$.

We will denote "N is a submodule of M_R" by $N \leq M_R$.

Example 1.6.2
Every ring R can be viewed both as a right R-module R_R and as a left R-module ${}_R R$ (Example 1.5.3). Submodules of R_R are exactly right ideals of R. Similarly, submodules of ${}_R R$ are exactly left ideals of R.

Submodules are the analog for modules of what subgroups are for groups. In groups, we also have the concepts of normal subgroups, which coincide with kernel of homomorphisms, and are the special subgroups with respect to which it is possible to construct quotient groups. Let us examine these concepts for the algebraic structures that should be known to the reader.

Semigroups. Semigroups are probably the simplest algebraic structures. A *semigroup* is a set S endowed with a binary operation $\cdot$, which is associative: $a(bc) = (ab)c$ for every $a, b, c \in S$. A *subsemigroup* T of a semigroup S is a subset of T that is multiplicatively closed: $a \in T$ and $b \in T$ imply $ab \in T$, where ab denotes the product of a and b in S. A *homomorphism* of the semigroup S into the semigroup S' is a mapping $\varphi: S \to S'$ such that $\varphi(ab) = \varphi(a)\varphi(b)$ for every $a, b \in S$. A *congruence* on a semigroup S is an equivalence relation $\sim$ on the set S (that is, a relation that is reflexive, symmetric and transitive) such that, for every $a, b, c, d \in S$, $a \sim b$ and $c \sim d$ imply $ac \sim bd$. This means that with a congruence

1.6 Submodules and Quotients

$\sim$ is "compatible with the operation", that is, "one can multiply term by term the two members of two equations":

$$a \sim b \text{ and } c \sim d \text{ imply } ac \sim bd.$$

For any semigroup homomorphism $\varphi \colon S \to S'$, the *kernel* of φ is the congruence $\sim_\varphi$ on S defined, for every $a, b \in S$, by $a \sim_\varphi b$ if $\varphi(a) = \varphi(b)$. Given a congruence $\sim$ on a semigroup S, the operation $\cdot$ on the quotient set $S/\sim$, set of all congruences classes $[s]_\sim = \{ x \in S \mid x \sim s \}$ when s ranges in S, defined by

$$[s]_\sim [s']_\sim = [ss']_\sim \qquad \text{for every } s, s' \in S$$

is well defined. The quotient set

$$S/\sim \; = \{ [s]_\sim \mid s \in S \}$$

is a semigroup. Every congruence on a subgroup is the kernel of a semigroup homomorphism: given a congruence $\sim$ on a semigroup S, $\sim$ turns out to be the kernel of the canonical projection $S \to S/\sim$.

Monoids. A *monoid* is a semigroup S with an identity: there exists an element $1_S \in S$ such that $a 1_S = 1_S a = a$ for every $a \in S$. A *submonoid* T of a monoid S is a subsemigroup of T such that $1_S \in T$. A *monoid homomorphism* of the monoid S into the monoid S' is a semigroup homomorphism $\varphi \colon S \to S'$ such that $\varphi(1_S) = 1_{S'}$. A *congruence* on a monoid S is the same as a congruence on the semigroup S. Since monoid homomorphisms are special semigroup homomorphisms, it follows that the *kernel* of a monoid homomorphisms $\varphi \colon S \to S'$ is the congruence $\sim_\varphi$ on the monoid S. Given a congruence $\sim$ on a monoid S, the quotient semigroup $S/\sim$ turns out to be a monoid. Every congruence on a monoid is the kernel of a monoid homomorphism.

Groups. A *group* is a monoid G with identity 1_G in which every element is invertible: for every $x \in G$, there exists an element $y \in G$ such that $xy = yx = 1_G$. Under these hypotheses, the *inverse* y in G of an element $x \in G$ is unique, and is denoted by x^{-1}. A *subgroup* H of a group G is a submonoid of G such that $x^{-1} \in H$ for every $x \in H$. Here x^{-1} denotes the inverse in G of the element $x \in H$. A *group homomorphism* of a group G into a group G' is a semigroup homomorphism $\varphi \colon G \to G'$ (One verifies that a semigroup homomorphism $\varphi \colon G \to G'$ automatically respect the identity, inverses, and powers of an element of G). A *congruence* on a group G is the same as a congruence on the semigroup G, but in this case there is a one-to-one correspondence between the set of all congruences on the group G and the set of all *normal* subgroups of G, that is, the subgroups H of G such that $xhx^{-1} \in G$ for every $x \in G, h \in H$. For every congruence $\sim$ on G, the corresponding normal subgroup of G is the set of all elements $x \in G$ with $x \sim 1_G$. Vice versa, for every normal subgroup H of G the corresponding congruence $\sim_H$ on G is defined, for every $x, y \in G$, by $x \sim_H y$ if $xy^{-1} \in H$. Thus congruences correspond, for groups, to normal subgroups. Of course, the kernel of a group homomorphism $\varphi \colon G \to G'$, as usually defined in group theory, is exactly the normal subgroup of G corresponding to the congruence that is the kernel of φ as defined for semigroups. Also, for every normal subgroup H of G, the quotient semigroup $G/\sim_H$ is exactly the usual quotient group G/H. Every normal subgroup is the kernel of a group homomorphism.

Rings. For rings, substructures are obviously subrings. As rings have two operations $+, \cdot$, congruences must be compatible with both operations. Thus a *congruence* on a ring R is an equivalence $\sim$ on the set R such that, for every $x, y, z, w \in R$, $x \sim y$ and $z \sim w$ imply $x + z \sim y + w$ and $xz \sim yw$ (i.e., "we can add and multiply term by term the two members of two equations"). For a ring R, there is a one-to-one correspondence between the set of all congruences on R and the set of all two-sided ideals of R. For every congruence $\sim$ on R, the corresponding two-sided ideal of R is the set of all $x \in R$ with $x \sim 0_R$. Vice versa, for every $I \trianglelefteq R$ the corresponding congruence $\sim_I$ on R is defined, for every $x, y \in R$, by $x \sim_I y$ if $x - y \in I$. Thus congruences correspond, for rings, to two-sided ideals. The kernel of a ring homomorphism $\varphi \colon R \to R'$, as usually defined in ring theory, is the two-sided ideal of R corresponding to the congruence $\sim$ on R defined, for every $x, y \in R$, by $x \sim y$ if $\varphi(x) = \varphi(y)$. Also, for every two-sided ideal I of R, the quotient ring R/I is the quotient set $R/\sim_I$ with the operations induced from those of R. Every two-sided ideal is the kernel of a ring homomorphism.

Right R-modules. Let R be a fixed ring and M_R a right R-module. The substructures of M_R are its submodules. A *congruence* on M_R is an equivalence $\sim$ on the set M_R such that, for every $x, y, z, w \in M_R$ and $r \in R$, $x \sim y$ and $z \sim w$ imply $x + z \sim y + w$ and $xr \sim yr$. There is a one-to-one correspondence between the set of all congruences of M_R and the set of all submodules of M_R. For every congruence $\sim$ on M_R, the corresponding submodule of M_R is the set of all $x \in M_R$ with $x \sim 0$. Vice versa, for every $N \leq M_R$ the corresponding congruence $\sim_N$ on M_R is defined, for every $x, y \in M_R$, by $x \sim_N y$ if $x - y \in N$. Thus congruences correspond to submodules. If $\varphi \colon M_R \to M'_R$ is a right R-module morphism, the *kernel* of φ is $\ker \varphi := \{ x \in M_R \mid \varphi(x) = 0 \}$. It is a submodule of M_R. Also for modules, the kernel $\ker \varphi$ of a homomorphism $\varphi \colon M_R \to M'_R$ is the submodule of M_R corresponding to the congruence $\sim$ on M_R defined, for every $x, y \in M_R$, by $x \sim y$ if $\varphi(x) = \varphi(y)$. If N is a submodule of M_R, the *quotient module* M_R/N is the set of all cosets $x + N$, where x ranges in M_R. The operations on M_R/N are defined by $(x + N) + (y + N) = (x + y) + N$ and $(x + N)r = xr + N$ for every $x, y \in N, r \in R$. These two operations turn out to be well defined, and M_R/N with these operations becomes a right R-module. The quotient module M_R/N is the quotient set $M_R/\sim_N$. The *canonical projection* $\pi \colon M_R \to M_R/N$ is defined by $\pi(x) = x + N$ for every $x \in M_R$. It is a right R-module morphism and its kernel is N. Hence submodules of M_R are exactly the kernels of the morphisms of M_R into some other right R-module.

1.7 Natural Transformations of Functors

Definition 1.7.1

A morphism $f\colon C \to C'$ in a category $\mathcal{C}$ is called an *isomorphism* if there exists a morphism $g\colon C' \to C$ in $\mathcal{C}$ such that $g \circ f = 1_C$ and $f \circ g = 1_{C'}$. In this case, the two objects C and C' of $\mathcal{C}$ are said to be *isomorphic*.

Examples 1.7.2
(1) In the categories **Set**, **Grp**, **Rng**, k-**Vect**, Mod-R, **Ab**, objects are (particular) sets, morphisms are mappings and the composition is the usual composition of mappings. In these categories, a morphism is an isomorphism if and only if it is a bijective mapping.
(2) In the category **Top**, the morphisms of a topological space X into a topological space Y are exactly all continuous mappings $X \to Y$. In this category, the isomorphisms $X \to Y$ are exactly the homeomorphisms $X \to Y$ (recall that, by definition, a homeomorphism $f\colon X \to Y$ between two topological spaces X and Y is a bijection f for which both f and its inverse f^{-1} are continuous mappings). Notice that the set of all homeomorphisms $X \to Y$ is, in general, properly contained in the set of all bijective continuous mappings $X \to Y$. For instance, if X is the topological space of the real numbers with the discrete topology and Y is the topological space of the real numbers with the usual topology, then the identity mapping $\iota\colon X \to Y$ is a continuous mapping, it is bijective, but it is not a homeomorphism. Hence it is a morphism, it is bijective, but it is not an isomorphism.
(3) Let **POSet** be the category of all partially ordered sets. In this category: the objects are all partially ordered sets; if $A, B \in$ Ob(**POSet**), the morphisms $A \to B$ are all increasing mappings of A into B, that is, $\mathrm{Hom}_{\mathbf{POSet}}(A, B) := \{ f\colon A \to B \mid \text{ for all } a, a' \in A,\ a \le a'$ implies $f(a) \le f(a') \}$. Hence two objects A and B in the category **POSet** are isomorphic if and only if there is a mapping $f\colon A \to B$ that is increasing, bijective and whose inverse mapping is also increasing, that is, $f\colon A \to B$ is bijective and for all $a, a' \in A$, $a \le a'$ if and only if $f(a) \le f(a')$. This is a stronger condition than being bijective and increasing only. For instance, assume that $A = B$ are both the partially ordered sets with two elements the two natural numbers 0 and 1. Suppose that B has the partial order induced by the usual order of natural numbers and A has the partial order defined by the equality $=$. Then the identity mapping $\iota\colon A \to B$ is an increasing mapping, it is bijective, but its inverse is not an increasing mapping. Hence ι is a morphism, it is bijective, but it is not an isomorphism in the category **POSet**.

Definition 1.7.3

Let $\mathcal{C}, \mathcal{D}$ be categories and $F, G\colon \mathcal{C} \to \mathcal{D}$ be covariant functors. A *natural transformation* $\eta\colon F \to G$ assigns to each object $C \in \mathrm{Ob}\,\mathcal{C}$ a morphism $\eta_C\colon F(C) \to G(C)$ such that, for every morphism $f\colon C \to C'$ in the category $\mathcal{C}$, the diagram

$$\begin{array}{ccc} F(C) & \xrightarrow{F(f)} & F(C') \\ \eta_C \downarrow & & \downarrow \eta_{C'} \\ G(C) & \xrightarrow[G(f)]{} & G(C') \end{array}$$

(1.3)

commutes.

Definition 1.7.4

A natural transformation $\eta\colon F \to G$ is called a *natural isomorphism* (or an *isomorphism of functors*, or a *functorial isomorphism*) if $\eta_C\colon F(C) \to G(C)$ is an isomorphism for every $C \in \mathrm{Ob}\,\mathcal{C}$.

Example 1.7.5
Let **CRng** be the category of all commutative rings and **Mon** be the category of monoids (the objects are all monoids, whose operation will be denoted by multiplication, and morphisms are all monoid homomorphisms). There is a covariant functor

$$F\colon \mathbf{CRng} \to \mathbf{Mon},$$

(a *forgetful functor*) that associates to any commutative ring $(R, +, \cdot)$ its multiplicative monoid $(R, \cdot)$, and that is the identity on morphisms. Fix an integer $n \geq 1$. There is a functor $\mathbb{M}_n\colon \mathbf{CRng} \to \mathbf{Mon}$ that assigns to every commutative ring R the monoid $\mathbb{M}_n(R)$ of all $n \times n$ matrices with entries in R. Here the operation on $\mathbb{M}_n(R)$ is the usual row by column multiplication of matrices. The functor $\mathbb{M}_n$ sends a morphism $\varphi\colon R \to S$ to the morphism $\mathbb{M}_n(\varphi)\colon \mathbb{M}_n(R) \to \mathbb{M}_n(S)$, $\mathbb{M}_n(\varphi)\colon A = (a_{ij})_{i,j} \mapsto \varphi(A) = (\varphi(a_{ij}))_{i,j}$. There is a natural transformation of functors $\det\colon \mathbb{M}_n \to F$, called *determinant*, that associates to each commutative ring R the monoid morphism $\det_R\colon \mathbb{M}_n(R) \to F(R)$, where $\det_R(A)$ denotes the determinant of A for every matrix $A \in \mathbb{M}_n(R)$. To prove it, fix a ring morphism $\varphi\colon R \to S$. We must show that the diagram

$$\begin{array}{ccc} \mathbb{M}_n(R) & \xrightarrow{\mathbb{M}_n(\varphi)} & \mathbb{M}_n(S) \\ {\scriptstyle \det_R}\downarrow & & \downarrow{\scriptstyle \det_S} \\ F(R) & \xrightarrow{F(\varphi)} & F(S) \end{array}$$

commutes. Take an arbitrary matrix $A = (a_{ij})_{i,j} \in \mathbb{M}_n(R)$. Then $\det_S \circ \mathbb{M}_n(\varphi)$ sends A to

$$(\det_S \circ \mathbb{M}_n(\varphi))(A) = \det_S(\varphi(a_{ij})_{i,j}) = \sum_{\sigma \in S_n} \mathrm{sgn}(\sigma)\varphi(a_{1\sigma(1)})\ldots\varphi(a_{n\sigma(n)}),$$

and $F(\varphi) \circ \det_R$ sends A to

$$(F(\varphi) \circ \det_R)(A) = \varphi(\det_R((a_{ij})_{i,j})) =$$
$$= \varphi\left(\sum_{\sigma \in S_n} \mathrm{sgn}(\sigma) a_{1\sigma(1)} \ldots a_{n\sigma(n)}\right) =$$
$$= \sum_{\sigma \in S_n} \mathrm{sgn}(\sigma)\varphi(a_{1\sigma(1)})\ldots\varphi(a_{n\sigma(n)}) = (\det_S \circ \mathbb{M}_n(S))(A).$$

1.7 Natural Transformations of Functors

Example 1.7.6
Fix a module M_R. Consider the covariant functor

$$\text{Hom}(M_R, -) \colon \text{Mod-}R \to \textbf{Ab}$$

that assigns to every module N_R the abelian group $\text{Hom}(M_R, N_R)$, and to every morphism $f \colon N_R \to N'_R$ the morphism

$$\text{Hom}(M_R, f) \colon \text{Hom}(M_R, N_R) \to \text{Hom}(M_R, N'_R),$$

defined by $\text{Hom}(M_R, f)(\alpha) = f \circ \alpha$ for every $\alpha \in \text{Hom}(M_R, N_R)$.

Let $g \colon M_R \to M'_R$ be a fixed right R-module homomorphism. There is a natural transformation of functors

$$\eta = \text{Hom}(g, -) \colon \text{Hom}(M'_R, -) \to \text{Hom}(M_R, -)$$

that assigns to each module N_R the abelian group morphism

$$\eta_{N_R} = \text{Hom}(g, N_R) \colon \text{Hom}(M'_R, N_R) \to \text{Hom}(M_R, N_R)$$

defined by $\text{Hom}(g, N_R)(\beta) = \beta \circ g$ for every $\beta \in \text{Hom}(M'_R, N_R)$. To see it, notice that, for every morphism $f \colon N_R \to N'_R$, the diagram

$$\begin{array}{ccc} \text{Hom}(M'_R, N_R) & \xrightarrow{\text{Hom}(M'_R, f) = f \circ -} & \text{Hom}(M'_R, N'_R) \\ {\scriptstyle \text{Hom}(g, N_R) = - \circ g} \downarrow & & \downarrow {\scriptstyle \text{Hom}(g, N'_R) = - \circ g} \\ \text{Hom}(M_R, N_R) & \xrightarrow{\text{Hom}(M_R, f) = f \circ -} & \text{Hom}(M_R, N'_R) \end{array}$$

commutes, because $f \circ (\alpha \circ g) = (f \circ \alpha) \circ g$.

Example 1.7.7
Consider the covariant functor

$$\text{Hom}(R_R, -) \colon \text{Mod-}R \to \textbf{Ab}$$

and the forgetful functor

$$F \colon \text{Mod-}R \to \textbf{Ab}.$$

Let $\eta \colon \text{Hom}(R_R, -) \to F$ be the natural transformation that assigns to every right R-module C_R the morphism

$$\begin{aligned} \eta_C \colon \text{Hom}(R_R, C_R) & \to C \\ \alpha & \mapsto \alpha(1_R). \end{aligned}$$

We must check that for, every $f: C_R \to C'_R$, the diagram

$$\begin{array}{ccc} \mathrm{Hom}(R_R, C_R) & \xrightarrow{\mathrm{Hom}(R_R, f)} & \mathrm{Hom}(R_R, C'_R) \\ \eta_C \downarrow & & \downarrow \eta_{C'} \\ C & \xrightarrow{f} & C' \end{array}$$

commutes. That is, for every $\alpha \in \mathrm{Hom}(R_R, C_R)$, $f(\alpha(1_R)) = (f \circ \alpha)(1_R)$, which is true.

For every right R-module C_R, the morphism η_C is a group isomorphism: its inverse is

$$\begin{aligned} \mu_C: C &\to \mathrm{Hom}(R_R, C_R) \\ c &\mapsto \mu_C(c), \end{aligned}$$

where $\mu_C(c)$ is the right R-module homomorphism that sends an element $r \in R$ to cr.

Example 1.7.8
Let k be a division ring and V_k a right vector space over k. The *dual* of V_k is $(V_k)^* = \mathrm{Hom}(V_k, {}_k k_k)$, which is a left k-module (Example 1.5.8), that is, a left vector space over k. Dually, if ${}_k V$ is a left vector space, its dual is $({}_k V)^* = \mathrm{Hom}({}_k V, {}_k k_k)$, which is a right vector space over k.

From now on, we will denote by **Vect**-k (k-**Vect**) the category of right (left, respectively) vector spaces over k and by **vect**-k (k-**vect**) the category of right (left) finite-dimensional vector spaces over k.

We can define the duality functors

$$\begin{array}{cccc} D = \mathrm{Hom}(-, {}_k k_k): & \mathbf{Vect}\text{--}k & \to & k\text{--}\mathbf{Vect} \\ & V_k & \mapsto & {}_k(V^*) \\ & (f: V_k \to W_k) & \mapsto & (f^{tr} = f^*: {}_k W^* \to {}_k V^*) \end{array}$$

and

$$\begin{array}{cccc} D = \mathrm{Hom}(-, {}_k k_k): & k\text{--}\mathbf{Vect} & \to & \mathbf{Vect}\text{--}k \\ & {}_k V & \mapsto & (V^*)_k \\ & (f: {}_k V \to {}_k W) & \mapsto & (f^{tr} = f^*: W_k^* \to V_k^*). \end{array}$$

(We are using the same notation D to indicate these two different functors, but this will cause no difficulty.) Both these D's are contravariant functors, so their composition is a covariant functor

$$\begin{array}{cccc} D \circ D: & \mathbf{Vect}\text{--}k & \to & \mathbf{Vect}\text{--}k \\ & V_k & \mapsto & (V^{**})_k \\ & (f: V_k \to W_k) & \mapsto & (f^{**}: V_k^{**} \to W_k^{**}), \end{array}$$

where

$$f^{**}(\varphi)(\alpha') = \varphi(\alpha' \circ f) \quad \text{for every } \varphi \in V^{**}, \alpha' \in W^*.$$

There is a natural transformation $\eta: I_{\mathbf{Vect}-k} \to D \circ D$ from the identity functor to the biduality functor given, for every right vector space V, by

$$\eta_V: V \to V^{**}$$

$$\eta_V(v)(\alpha) = \alpha(v) \quad \text{for every } v \in V, \alpha \in V^*.$$

1.7 Natural Transformations of Functors

It is a natural transformation, because, for every k-linear transformation $f: V_k \to W_k$, the following diagram commutes.

$$\begin{array}{ccc} V_k & \xrightarrow{f} & W_k \\ \eta_V \downarrow & & \downarrow \eta_W \\ V_k^{**} & \xrightarrow{f^{**}} & W_k^{**} \end{array}$$

If we consider the category **vect**-k, then we know that each η_V is an isomorphism, so η is a natural isomorphism between I and $D \circ D$ viewed as functors of **vect**-k into **vect**-k.

Exercise 1.7.9
For every group $(G, \cdot)$, it is possible to define the *opposite group* $(G, \circ)$, where $\circ: G \times G \to G$ is the operation on the set G defined by $g \circ g' = g' \cdot g$ for every $g, g' \in G$ (Cf. Exercise 1.4.7). The opposite group of a group G will be denoted by G^{op}. Show that every group is naturally isomorphic to its opposite group. More precisely, let Op: **Grp** $\to$ **Grp** be the functor that maps every group G to its opposite group G^{op}, and every morphism f in **Grp** to f (Show that Op is a functor!) Let I: **Grp** $\to$ **Grp** be the be the identity functor. Show that there is a natural isomorphism $\eta: I \to \text{Op}$, where, for every group G, $\eta_G: I(G) = G \to \text{Op}(G) = G^{\text{op}}$ is defined by $\eta_G(g) = g^{-1}$ for every $g \in G$.

Notice that, for a ring R, the ring R is not necessarily isomorphic to its opposite ring R^{op}. For instance, we will give in Exercise 2.11.7 an example of a ring R that is left noetherian, but not right noetherian. Such a ring cannot be isomorphic to its opposite ring.

If $\mathcal{C}, \mathcal{D}$ are categories and $F, G: \mathcal{C} \to \mathcal{D}$ are *contravariant* functors, a natural transformation $\eta: F \to G$ is defined as in Definition 1.7.4, the only difference being that the diagram that must commute for every morphism $f: C \to C'$ is now

$$\begin{array}{ccc} F(C') & \xrightarrow{F(f)} & F(C) \\ \eta_{C'} \downarrow & & \downarrow \eta_C \\ G(C') & \xrightarrow{G(f)} & G(C), \end{array} \quad (1.4)$$

instead of Diagram (1.3).

Exercise 1.7.10
Let R be a ring, Mod-R the category of all right R-modules and I: Mod-R $\to$ Mod-R the identity functor. Show that there is a one-to-one correspondence between the "set" of all natural transformations $\eta: I \to I$ and the center $Z(R)$ of the ring R.

[Hint: Consider the right R-module morphism $\eta_R: R_R \to R_R$, determining a unique element $r \in R$ such that η_R is left multiplication λ_r by r. Then show that $r \in Z(R)$, so that λ_r coincides with right multiplication ρ_r by r. Then consider, for any right R-module M_R and any element $m \in M_R$, the right R-module morphism $f: R_R \to M_R$, $1 \mapsto m$. Show that $\eta_{M_R}: M_R \to M_R$ is also right multiplication by r, for every right R-module M_R.]

1.8 Sums and Intersections, Direct Sums and Direct Products

Let M_R be an R-module and $\mathcal{F} = \{\, N_\lambda \mid \lambda \in \Lambda \,\}$ be a family of submodules of M_R. The *sum* $\sum_{\lambda \in \Lambda} N_\lambda := \{\, x_{\lambda_1} + \cdots + x_{\lambda_n} \mid n \geq 1,\ \lambda_1, \ldots, \lambda_n \in \Lambda,\ x_{\lambda_t} \in N_{\lambda_t}$ for every $t = 1, \ldots, n \,\}$ and the *intersection* $\bigcap_{\lambda \in \Lambda} N_\lambda$ are submodules of M_R.

If M_R, N_R are right R-modules, their *(external) direct sum* is the set of all pairs (x, y), with $x \in M_R$ and $y \in N_R$, with the operations defined componentwise:

$$(x, y) + (x', y') = (x + x', y + y'), \qquad (x, y)r = (xr, yr)$$

for every $x \in M_R$, $y \in N_R$, $r \in R$. Hence, as a set, the direct sum of M_R and N_R is their cartesian product $M_R \times N_R$. The direct sum of M_R and N_R will be denoted by $M_R \oplus N_R$.

More generally, given an arbitrary family $\{\, M_\lambda \mid \lambda \in \Lambda \,\}$ of right R-modules, it is possible to define the direct product of the family as follows. Consider the cartesian product $\prod_{\lambda \in \Lambda} M_\lambda$ of the family, that is, the set of all mappings $f \colon \Lambda \to \bigcup_{\lambda \in \Lambda} M_\lambda$ such that $f(\lambda) \in M_\lambda$ for every $\lambda \in \Lambda$. In $\prod_{\lambda \in \Lambda} M_\lambda$, we can define the operations by

$$(f + g)(\lambda) = f(\lambda) + g(\lambda), \qquad (fr)(\lambda) = f(\lambda)r$$

for every $f, g \in \prod_{\lambda \in \Lambda} M_\lambda$, $\lambda \in \Lambda$, $r \in R$. Then $\prod_{\lambda \in \Lambda} M_\lambda$, with these operations, becomes a right R-module, called the *direct product* of the modules M_λ, $\lambda \in \Lambda$. (More informally, we can denote the elements of the direct product $\prod_{\lambda \in \Lambda} M_\lambda$ by Λ-*tuples* $(\ldots, x_\lambda, \ldots)$. In this notation, the operations are defined by

$$(\ldots, x_\lambda, \ldots) + (\ldots, x'_\lambda, \ldots) = (\ldots, x_\lambda + x'_\lambda, \ldots)$$
$$(\ldots, x_\lambda, \ldots)r = (\ldots, x_\lambda r, \ldots)$$

for every $(\ldots, x_\lambda, \ldots), (\ldots, x'_\lambda, \ldots) \in \prod_{\lambda \in \Lambda} M_\lambda$, $r \in R$.)

The *(external) direct sum* $\bigoplus_{\lambda \in \Lambda} M_\lambda$ of the family $\{\, M_\lambda \mid \lambda \in \Lambda \,\}$ is the set of all elements of the direct product $\prod_{\lambda \in \Lambda} M_\lambda$ that are almost all zero:

$$\bigoplus_{\lambda \in \Lambda} M_\lambda := \{\, f \in \prod_{\lambda \in \Lambda} M_\lambda \mid f(\lambda) = 0_{M_\lambda} \text{ for almost all } \lambda \in \Lambda \,\}.$$

(If we denote the elements of $\prod_{\lambda \in \Lambda} M_\lambda$ by Λ-tuples, this definition becomes

$$\bigoplus_{\lambda \in \Lambda} M_\lambda := \{\, (\ldots, x_\lambda, \ldots) \in \prod_{\lambda \in \Lambda} M_\lambda \mid x_\lambda = 0_{M_\lambda} \text{ for almost all } \lambda \in \Lambda \,\}.)$$

Let us pass to consider the *internal* direct sum of submodules. Let M_R be an R-module and N, N' two submodules of M_R. If $N + N' = M_R$ and $N \cap N' = 0$, we say that M_R is the *internal direct sum* of its submodules N and N'. It is easily checked

1.8 Sums and Intersections, Direct Sums and Direct Products

that $N + N' = M_R$ and $N \cap N' = 0$ if and only if the mapping $\vartheta \colon N \oplus N' \to M_R$, defined by $\vartheta \colon (n, n') \mapsto n + n'$, is an isomorphism. Also, if N_R, N'_R and M_R are right R-modules and there is an isomorphism $\varphi \colon N_R \oplus N'_R \to M_R$, then M_R is the internal direct sum of its submodules $\varphi(N_R \oplus 0)$ and $\varphi(0 \oplus N'_R)$.

Exercise 1.8.1
Show that if a module N is a direct summand of a module M, then N is a direct summand of any module between N and M. More precisely, show that if $M = N \oplus N'$ and $N \leq P \leq M$, then $P = N \oplus (N' \cap P)$.

Lemma 1.8.2
Let $\mathcal{F} = \{ N_\lambda \mid \lambda \in \Lambda \}$ be a family of submodules of a right R-module M_R. The following conditions are equivalent:

(i) $\sum_{\lambda \in \Lambda} N_\lambda = M_R$ and, for every $\mu \in \Lambda$, $(\sum_{\lambda \in \Lambda \setminus \{\mu\}} N_\lambda) \cap N_\mu = 0$.
(ii) The mapping $\vartheta \colon \bigoplus_{\lambda \in \Lambda} N_\lambda \to M_R$, $\vartheta \colon (\ldots, x_\lambda, \ldots) \mapsto \sum_{\lambda \in \Lambda} x_\lambda$ is an isomorphism.

For any family $\mathcal{F} = \{ N_\lambda \mid \lambda \in \Lambda \}$ of modules and any $\mu \in \Lambda$, there are canonical homomorphisms

$$\varepsilon_\mu \colon N_\mu \to \bigoplus_{\lambda \in \Lambda} N_\lambda \quad \text{and} \quad \pi_\mu \colon \prod_{\lambda \in \Lambda} N_\lambda \to N_\mu.$$

Proposition 1.8.3
For every module M_R and every family $\mathcal{F} = \{ N_\lambda \mid \lambda \in \Lambda \}$ of right R-modules, there are:

(i) a canonical group isomorphism

$$Hom(M_R, \prod_{\lambda \in \Lambda} N_\lambda) \to \prod_{\lambda \in \Lambda} Hom(M_R, N_\lambda),$$

defined by $\varphi \in Hom(M_R, \prod_{\lambda \in \Lambda} N_\lambda) \mapsto (\pi_\lambda \circ \varphi)_{\lambda \in \Lambda}$; and

(ii) a canonical group isomorphism

$$Hom(\bigoplus_{\lambda \in \Lambda} N_\lambda, M_R) \to \prod_{\lambda \in \Lambda} Hom(N_\lambda, M_R),$$

defined by $\varphi \in Hom(\bigoplus_{\lambda \in \Lambda} N_\lambda, M_R) \mapsto (\varphi \circ \varepsilon_\lambda)_{\lambda \in \Lambda}$.

The proof is left as an exercise to the reader. Properties (i) and (ii) of Proposition 1.8.3 can also be stated as *Universal Properties*, in the following sense.

> **Proposition 1.8.4 (Universal Property of Direct Product)**
> Let $\{N_\lambda \mid \lambda \in \Lambda\}$ be a family of right R-modules, $\prod_{\lambda \in \Lambda} N_\lambda$ be its direct product and $\pi_\mu \colon \prod_{\lambda \in \Lambda} N_\lambda \to N_\mu$, $\mu \in \Lambda$, be the canonical projection. Then, for every right R-module M_R and every family $\{\varphi_\lambda \colon M_R \to N_\lambda \mid \lambda \in \Lambda\}$ of right R-module homomorphisms, there exists a unique right R-module homomorphism $\varphi \colon M_R \to \prod_{\lambda \in \Lambda} N_\lambda$ making all the diagrams
>
> $$\begin{array}{ccc} M_R & \xrightarrow{\varphi} & \prod_{\lambda \in \Lambda} N_\lambda, \\ & \searrow{\varphi_\mu} & \downarrow{\pi_\mu} \\ & & N_\mu \end{array}$$
>
> $\mu \in \Lambda$, commute.

> **Proposition 1.8.5 (Universal Property of Direct Sum)**
> Let $\{N_\lambda \mid \lambda \in \Lambda\}$ be a family of right R-modules, $\bigoplus_{\lambda \in \Lambda} N_\lambda$ be its direct sum and $\varepsilon_\mu \colon N_\mu \to \bigoplus_{\lambda \in \Lambda} N_\lambda$, $\mu \in \Lambda$, be the embeddings. Then, for every right R-module M_R and every family $\{\varphi_\lambda \colon N_\lambda \to M_R \mid \lambda \in \Lambda\}$ of right R-module homomorphisms, there exists a unique right R-module homomorphism $\varphi \colon \bigoplus_{\lambda \in \Lambda} N_\lambda \to M_R$ making all the diagrams
>
>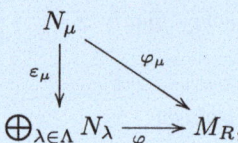
>
> $\mu \in \Lambda$, commute.

Homomorphisms Between Finite Direct Sums of Modules

Recall that every linear transformation $f \colon V \to W$ of an m-dimensional vector space V (with a fixed ordered basis $\{v_1, \ldots, v_m\}$) into an n-dimensional vector space W (with a fixed ordered basis $\{w_1, \ldots, w_n\}$) is completely described by an

1.8 Sums and Intersections, Direct Sums and Direct Products

$n \times m$ matrix, its associated matrix $A_f = (a_{ij})$, where $f(v_j) = \sum_{i=1}^{n} w_i a_{ij}$. This idea can be adapted to direct sums of finitely many modules.

Suppose $M_1, \ldots, M_m$ and $N_1, \ldots, N_n$ are right R-modules. By Proposition 1.8.3, we already know that

$$\mathrm{Hom}_R \left(\bigoplus_{i=1}^{m} M_i, \bigoplus_{j=1}^{n} N_j \right) \cong \bigoplus_{i=1}^{m} \bigoplus_{j=1}^{n} \mathrm{Hom}_R(M_i, N_j).$$

Set $M_R = M_1 \oplus \cdots \oplus M_m$ and $N_R = N_1 \oplus \cdots \oplus N_n$. Let $\varepsilon_j \colon M_j \to M$ denote the embedding for every $j = 1, \ldots, m$ and $\pi_i \colon N \to N_i$ denote the canonical projection corresponding to the direct-sum decomposition $N_R = N_1 \oplus \cdots \oplus N_n$. It is possible to write the elements of the group $\mathrm{Hom}_R(M_R, N_R)$ as $n \times m$ matrices instead of as nm-tuples, so that we get a group isomorphism

$$F \colon \mathrm{Hom}_R(M_R, N_R) \to \begin{pmatrix} \mathrm{Hom}_R(M_1, N_1) & \cdots & \mathrm{Hom}_R(M_m, N_1) \\ \mathrm{Hom}_R(M_1, N_2) & \cdots & \mathrm{Hom}_R(M_m, N_2) \\ \vdots & & \vdots \\ \mathrm{Hom}_R(M_1, N_n) & \cdots & \mathrm{Hom}_R(M_m, N_n) \end{pmatrix}. \tag{1.5}$$

This group morphism F maps a morphism $\varphi \colon M_R \to N_R$ to the matrix $(\varphi_{ij})_{ij}$, where $\varphi_{ij} = \pi_i \varphi \varepsilon_j$. Moreover, if $x = (x_1, \ldots, x_m) \in M_R$ is any element with $x_j \in M_j$, then

$$\varphi(x) = \begin{pmatrix} \varphi_{11} & \cdots & \varphi_{1m} \\ \varphi_{21} & \cdots & \varphi_{2m} \\ \vdots & & \vdots \\ \varphi_{n1} & \cdots & \varphi_{nm} \end{pmatrix} \begin{pmatrix} x_1 \\ x_2 \\ \vdots \\ x_m \end{pmatrix}.$$

Notice that if $R = k$ is a field and $V = v_1 k \oplus \cdots \oplus v_m k$ is a right k-vector space of dimension m and $W = w_1 k \oplus \cdots \oplus w_n k$ is a right k-vector space of dimension n, the isomorphism we've just set up corresponds to the usual isomorphism $\mathrm{Hom}_k(V, W) \cong M_{n \times m}(k)$, because $\mathrm{Hom}_k(v_j k, w_i k)$ is canonically isomorphic to k for all $i = 1, \ldots, n$ and $j = 1, \ldots, m$.

If $M_R = N_R = M_1 \oplus \cdots \oplus M_n$, the group isomorphism F becomes a ring isomorphism

$$F \colon \mathrm{End}_R(M_R) \to \begin{pmatrix} \mathrm{Hom}_R(M_1, M_1) & \cdots & \mathrm{Hom}_R(M_n, M_1) \\ \mathrm{Hom}_R(M_1, M_2) & \cdots & \mathrm{Hom}_R(M_n, M_2) \\ \vdots & & \vdots \\ \mathrm{Hom}_R(M_1, M_n) & \cdots & \mathrm{Hom}_R(M_n, M_n) \end{pmatrix}. \tag{1.6}$$

If we further assume that, in the direct-sum decomposition $M_1 \oplus \cdots \oplus M_n$ of $M_R = N_R$, the M_i's have the property that $\mathrm{Hom}(M_i, M_j) = 0$ for $i \neq j$, we get that

$$\mathrm{End}_R(M_R) \cong \begin{pmatrix} \mathrm{End}_R(M_1) & 0 & \cdots & 0 \\ 0 & \mathrm{End}_R(M_2) & & \\ \vdots & & \ddots & \\ 0 & 0 & & \mathrm{End}_R(M_n) \end{pmatrix}$$

is isomorphic to the direct product $\mathrm{End}_R(M_1) \times \cdots \times \mathrm{End}_R(M_n)$.

Finally, if we assume $M_i \cong M_j$ for all $i, j = 1, \ldots, n$, so that $M_R \cong M_1^n$, then

$$\mathrm{End}_R(M_R) \cong M_n(\mathrm{End}_R(M_1)), \tag{1.7}$$

the ring of $n \times n$ matrices with entries in $\mathrm{End}_R(M_1)$.

1.9 Isomorphism Theorems

The first two results of this Section can be easily proved by the reader:

Theorem 1.9.1 (First Isomorphism Theorem, or Fundamental Isomorphism Theorem)
Let $f: M_R \to M'_R$ be an R-module homomorphism. Let $\ker f$ be its kernel and $\pi: M_R \to M_R/\ker f$ the canonical projection. Then:

(i) *There exists a unique mapping $\widetilde{f}: M_R/\ker f \to M'_R$ making the diagram*

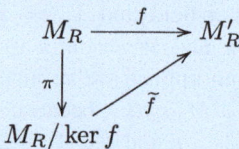

commute, that is, such that $\widetilde{f} \circ \pi = f$.
(ii) *The mapping $\widetilde{f}$ is an R-module monomorphism.*
(iii) *If f is onto, then $\widetilde{f}$ is an isomorphism.*

1.9 Isomorphism Theorems

Corollary 1.9.2
If $f: M_R \to M'_R$ is an R-module homomorphism, then
$$M_R/\ker f \cong f(M_R).$$

Theorem 1.9.3 (Correspondence Theorem)
Let $f: M_R \to M'_R$ be a right R-module homomorphism. Set $\mathcal{L} := \{N \mid N \leq M_R,\ N \supseteq \ker(f)\}$ and $\mathcal{L}' := \{N' \mid N' \leq M'_R,\ N' \subseteq f(M_R)\}$, and partially order $\mathcal{L}$ and $\mathcal{L}'$ by set inclusion. Then there is a partially ordered set isomorphism $\Phi: \mathcal{L} \to \mathcal{L}'$ defined by $\Phi(N) = f(N)$ for every $N \in \mathcal{L}$. The inverse $\Phi^{-1}: \mathcal{L}' \to \mathcal{L}$ of Φ is defined by $\Phi^{-1}(N') = f^{-1}(N')$ for every $N' \in \mathcal{L}'$.

To prove the Correspondence Theorem, notice that for every submodule N of M_R one has $f^{-1}(f(N)) = N + \ker f$, so that for every $N \in \mathcal{L}$ one has $f^{-1}(f(N)) = N$. Similarly, for every submodule N' of M'_R one has $f(f^{-1}(N')) = N' \cap f(M)$, so that $f(f^{-1}(N')) = N'$ for every $N' \in \mathcal{L}'$. Thus the positions $N \in \mathcal{L} \mapsto f(N)$ and $N' \in \mathcal{L}' \mapsto f^{-1}(N')$ are one the inverse of the other.

Theorem 1.9.4 (Second Isomorphism Theorem)
Let M_R be an R-module and let A, B be submodules of M_R. Then $A/(A \cap B) \cong (A + B)/B$.

Proof Consider the mapping $f: A \to (A+B)/B$ defined by $f(a) = a + B$ for every $a \in A$. It is easily seen that f is an R-module epimorphism. Its kernel is $\ker f = \{a \in A \mid a + B = B\} = \{a \in A \mid a \in B\} = A \cap B$. From the First Isomorphism Theorem, we get that $A/(A \cap B) \cong (A + B)/B$. □

Theorem 1.9.5 (Third Isomorphism Theorem)
Let M be an R-module and let $A \supseteq B$ be submodules of M_R. Then $\dfrac{M/B}{A/B} \cong M/A$.

Proof Consider the mapping $f: M/B \to M/A$ defined by $f(x+B) = x+A$ for every $x \in M$. It is easily seen that f is a well defined right R-module epimorphism. It kernel is

$$\ker f = \{x + B \mid x \in M, \ x + A = A\} = \{x + B \mid x \in A\} = A/B.$$

From the First Isomorphism Theorem, we get that $\dfrac{M/B}{A/B} \cong M/A$. □

1.10 Sets of Generators, Maximal Submodules

Let M_R be a right R-module and X a subset of M_R. Let $\mathcal{F}$ be the family of all the submodules of M_R that contain X. The family $\mathcal{F}$ is always non-empty, because, for instance, it contains M_R itself. The intersection of all the submodules in $\mathcal{F}$ is the smallest submodule of M_R that contains X. It is called the submodule of M_R *generated* by X and it is denoted by XR (or $\langle X \rangle$). It is easily seen that if $X = \emptyset$, then XR is the zero submodule of M_R. If $X \neq \emptyset$, then $XR = \{x_1 r_1 + \cdots + x_n r_n \mid n \geq 1, x_i \in X \text{ and } r_i \in R \text{ for } i = 1, \ldots, n\}$.

We say that a subset X of a right R-module M_R is a *set of generators* of M_R if $XR = M_R$. Except for the trivial case of $M_R = 0$, in which every subset of M_R is a set of generators, we have that a subset X of a module M_R is a set of generators for M_R if and only if for every $m \in M_R$ there exists finitely many elements $x_1, \ldots, x_n \in X$ and $r_1, \ldots, r_n \in R$ with $m = x_1 r_1 + \cdots + x_n r_n$.

A module M_R is *finitely generated* if it has a finite set of generators. A module M_R is *cyclic* if it has a set of generators with one element.

Lemma 1.10.1
A module M_R is cyclic if and only if it is isomorphic to R_R/I for some right ideal I of R.

Proof If M_R is cyclic and x is a generator of M_R, apply the First Isomorphism Theorem 1.9.1 to the module epimorphism $f: R_R \to M_R$ defined by $f(r) = xr$ for every $r \in R_R$. Then $R_R/I \cong M_R$ where $I := \ker(f)$.

For the converse, notice that R_R/I is cyclic generated by $1 + I$, and that every module isomorphic to a cyclic module is a cyclic module. □

A submodule N of a right module M_R is *proper* if it is different from M_R.

A *maximal submodule* N of M_R is a submodule of M_R that is maximal in the set of all proper submodules of M_R partially ordered by set inclusion. Thus a submodule N of M_R is maximal if $N < M_R$ and, for every submodule P of M_R with $N \leq P$, either $P = N$ or $P = M_R$.

1.10 Sets of Generators, Maximal Submodules

Proposition 1.10.2
Every non-zero finitely generated module has (at least) a maximal submodule. More generally, every proper submodule of a finitely generated module M_R is contained in a maximal submodule of M_R.

Proof Fix $K < M_R$. As M_R is finitely generated, there exist $x_1, \ldots, x_n \in M$ such that $M = K + x_1 R + \ldots + x_n R$. Choose one such set $\{x_1, \ldots, x_n\}$ with n as small as possible. Notice that $n \geq 1$ because K is proper. Hence $L := K + x_2 R + \ldots + x_n R < M_R$, and $x_1 \notin L$. Consider the set $\mathcal{P} = \{ N \mid L \leq N < M_R \}$. Notice that $\mathcal{P}$ is non-empty, because $L \in \mathcal{P}$. Partially order $\mathcal{P}$ by set inclusion. In order to apply Zorn's Lemma, we must show that every chain in $\mathcal{P}$ has an upper bound in $\mathcal{P}$. Fix a chain $\mathcal{C} = \{ N_\lambda \mid \lambda \in \Lambda \}$ in $\mathcal{P}$. Set $N^* := \bigcup_{\lambda \in \Lambda} N_\lambda$. It is easily seen that N^* is a submodule of M, which clearly contains L. Since $L \leq N_\lambda < M_R$ for every $\lambda \in \Lambda$, we have that x_1 does not belong to each N_λ, so that $x_1 \notin N^*$. Thus $N^* \in \mathcal{P}$. Now, clearly N^* is an upper bound of the chain $\mathcal{C}$, so that by Zorn's lemma $\mathcal{P}$ has a maximal element. Also, any maximal element of $\mathcal{P}$ is a maximal submodule of M_R. □

Exercise 1.10.3
Let M_R be a right R-module and

$$\mathcal{F} = \{ N_\lambda \mid \lambda \in \Lambda \}$$

be a family of right R-modules. Since

$$\bigoplus_{\lambda \in \Lambda} N_\lambda$$

is a submodule of

$$\prod_{\lambda \in \Lambda} N_\lambda,$$

we have that $\mathrm{Hom}(M_R, \bigoplus_{\lambda \in \Lambda} N_\lambda)$ is a subgroup of $\mathrm{Hom}(M_R, \prod_{\lambda \in \Lambda} N_\lambda)$. Also,

$$\bigoplus_{\lambda \in \Lambda} \mathrm{Hom}(N_\lambda, M_R)$$

is a subgroup of

$$\prod_{\lambda \in \Lambda} \mathrm{Hom}(N_\lambda, M_R).$$

Show that if M_R is finitely generated, the canonical isomorphism

$$\mathrm{Hom}(M_R, \prod_{\lambda \in \Lambda} N_\lambda) \to \prod_{\lambda \in \Lambda} \mathrm{Hom}(M_R, N_\lambda)$$

of Proposition 1.8.3(i) induces by restriction an isomorphism

$$\text{Hom}(M_R, \bigoplus_{\lambda \in \Lambda} N_\lambda) \to \bigoplus_{\lambda \in \Lambda} \text{Hom}(M_R, N_\lambda).$$

1.11 The Lattice of Submodules

Let $(A, \leq)$ be a partially ordered set and B a subset of A. Recall that an *upper bound of* B (or an *upper bound of* B *in* A) is any element $a \in A$ such that $b \leq a$ for every $b \in B$. Similarly, a *lower bound of* B is any element $a \in A$ such that $a \leq b$ for every $b \in B$. The *least upper bound* of B (in A) is the least element in the set of all $a \in A$ that are upper bounds of B. That is, $a \in A$ is the least upper bound of B in A if and only if $b \leq a$ for every $b \in B$, and, for every $x \in A$, if $b \leq x$ for every $b \in B$, then $a \leq x$. Similarly for the *greatest lower bound*. The least upper bound and the greatest lower bound of a subset B do not exist in general, but when they exist, they are unique.

A *lattice* is a partially ordered set $(L, \leq)$ such that, for every $x, y \in L$, the subset $\{x, y\}$ of L has a least upper bound and a greatest lower bound. As least upper bounds and greatest lower bounds are unique, the least upper bound of $\{x, y\}$ will be denoted $x \vee y$, and the greatest lower bound will be denoted $x \wedge y$. Thus $x \vee y = \min\{z \in L \mid x \leq z, y \leq z\}$.

In the next proposition, we will show that lattices, which we have defined as particular partially ordered sets, can also be introduced as algebraic structures, that is, as sets endowed with two binary operations satisfying suitable axioms.

Proposition 1.11.1
Let $(L, \leq)$ be a lattice. Define two operations in the set L by $\vee \colon L \times L \to L$, $(x, y) \mapsto x \vee y$ for every $x, y \in L$, and $\wedge \colon L \times L \to L$, $(x, y) \mapsto x \wedge y$ for every $x, y \in L$. Then the operations $\vee$ and $\wedge$ satisfy the following properties:

(i) *commutativity: $x \vee y = y \vee x$, $x \wedge y = y \wedge x$ for every $x, y \in L$;*
(ii) *associativity: $x \vee (y \vee z) = (x \vee y) \vee z$, $x \wedge (y \wedge z) = (x \wedge y) \wedge z$ for every $x, y, z \in L$;*
(iii) *absorption law: $x \vee (x \wedge y) = x$, $x \wedge (x \vee y) = x$ for every $x, y \in L$.*

Conversely, let $(L, \vee, \wedge)$ be a set L with two operations $\vee$ and $\wedge$ satisfying the three properties (i), (ii), (iii) above. Define a relation $\leq$ in the set L by setting, for every $x, y \in L$, $x \leq y$ if $x \wedge y = x$. Then $(L, \leq)$ is a lattice.

If L, L' are lattices, a mapping $\varphi \colon L \to L'$ is a *lattice homomorphism* if $\varphi(x \vee y) = \varphi(x) \vee \varphi(y)$ and $\varphi(x \wedge y) = \varphi(x) \wedge \varphi(y)$ for every $x, y \in L$. Every lattice

1.11 The Lattice of Submodules

homomorphism $\varphi\colon L \to L'$ is an increasing mapping, that is, for every $x, y \in L$, $x \le y$ implies $\varphi(x) \le \varphi(y)$, but not conversely. Obviously, lattices form a category. (More precisely, we have two isomorphic categories, one whose objects are the lattices defined as partially ordered sets and the other with objects the lattices defined as algebraic structures with two binary operations $\vee$ and $\wedge$.) It is easily checked that a mapping $\varphi\colon L \to L$ is an isomorphism in the category of lattices if and only if it is a bijection and for every $x, y \in L$, $x \le y \Leftrightarrow \varphi(x) \le \varphi(y)$, if and only if it is a bijection and $\varphi(x \vee y) = \varphi(x) \vee \varphi(y)$, $\varphi(x \wedge y) = \varphi(x) \wedge \varphi(y)$ for every $x, y \in L$.

Our main example of lattice will be, for any module M_R, *the lattice $\mathcal{L}(M_R)$ of all submodules of M_R*. Its elements are the submodules of M_R. The partial order is set inclusion. Then $(\mathcal{L}(M_R), \subseteq)$ is a lattice, where, for every $N, N' \in \mathcal{L}(M_R)$, $N \vee N' = N + N'$ and $N \wedge N' = N \cap N'$. A submodule of M_R is maximal if and only if it is a maximal element in the partially ordered set $\mathcal{L}(M_R) \setminus \{M_R\}$.

Exercise 1.11.2 (The Prüfer Group)

Let p be a prime. Consider the set $M = \left\{ \dfrac{a}{p^n} \mid a, n \in \mathbb{Z},\ n \ge 0 \right\} \subseteq \mathbb{Q}$. Show that M is a subgroup of $\mathbb{Q}$ that contains $\mathbb{Z}$. The group $\mathbb{Z}(p^\infty) := M/\mathbb{Z}$ is called the *Prüfer group*. Prove that:

 (i) The additive group $\mathbb{Z}(p^\infty)$ is *divisible*, that is, for every $x \in \mathbb{Z}(p^\infty)$ and every integer $m > 0$ there exists $y \in \mathbb{Z}(p^\infty)$ such that $x = my$.
 (ii) Every proper subgroup of $\mathbb{Z}(p^\infty)$ is cyclic and generated by $\dfrac{1}{p^n} + \mathbb{Z}$ for some $n \ge 0$.
 (iii) The cyclic subgroup of $\mathbb{Z}(p^\infty)$ generated by $\dfrac{1}{p^n} + \mathbb{Z}$ has order p^n, hence is isomorphic to $\mathbb{Z}/p^n\mathbb{Z}$.
 (iv) The lattices $\mathcal{L}(\mathbb{Z}(p^\infty))$ and $\mathbb{N} \cup \{+\infty\}$ are isomorphic.
 (v) The $\mathbb{Z}$-module $\mathbb{Z}(p^\infty)_\mathbb{Z}$ does not have maximal submodules.
 (vi) The abelian group $\mathbb{Z}(p^\infty)$ is isomorphic to the subgroup of the multiplicative group $\mathbb{C}^* = \mathbb{C} \setminus \{0\}$ consisting of all complex number z that are p^n-th roots of unity for some $n \ge 0$, that is, all $z \in \mathbb{C}^*$ such that there exists $n \ge 0$ with $z^{p^n} = 1$.
 (vii) Show that a subset X of $\mathbb{Z}(p^\infty)$ is a set of generators for $\mathbb{Z}(p^\infty)$ if and only if it is infinite. In particular, every set of generators for $\mathbb{Z}(p^\infty)$ properly contains a set of generators for $\mathbb{Z}(p^\infty)$, so that $\mathbb{Z}(p^\infty)$ does not have minimal sets of generators.

Some Classes of Modules

2.1 Free Modules

We have concluded the presentation of the basic notions about rings and modules. We will now begin the study of some of the most important classes of modules: the classes of free, projective, simple, semisimple modules, and so on. For the modules in each class we will give the definition, some characterizations, the main properties, we will see how the modules decompose into direct sums, whether these direct-sum decompositions are essentially unique, etc.

We have already seen in Sect. 1.10 what a set of generators of a module is. Let us see what a *free* set of generators is.

> **Definition 2.1.1**
>
> Let X be a set of generators of a right R-module M_R. The set X is called a *free* set of generators if, for every $n \geq 1$, $x_1, \ldots, x_n$ distinct elements of X and $r_1, \ldots, r_n$ in R, one has that $x_1 \cdot r_1 + \ldots + x_n \cdot r_n = 0$ implies $r_1 = \ldots = r_n = 0_R$. The empty set is a free set of generators of the zero module.

Notice that every module M_R has sets X of generators, for instance $X = M_R$. Not every module has free sets of generators. For instance, the $\mathbb{Z}$-module $\mathbb{Z}/n\mathbb{Z}$ does not have a free set of generators for $n \geq 2$.

> **Definition 2.1.2**
>
> A right R-module M_R is said to be *free* if it has a free set of generators.

Let M_R be a right R-module and X a subset of M_R. We know that X is a set of generators of M_R if and only if every element of M_R can be written as a linear combination of elements of X. It is easily seen that X is a free set of generators of

M_R if and only if every element of M_R can be written as a linear combination of distinct elements of X in a unique way; that is, $x_1 \cdot r_1 + \ldots + x_n \cdot r_n = x_1 \cdot r_1' + \ldots + x_n \cdot r_n'$ implies $r_1 = r_1', \ldots, r_n = r_n'$.

Example 2.1.3
Let R be a ring and let X be an arbitrary set. Let $R^{(X)}$ be the set of all mappings $f: X \to R$ such that $f(x) = 0$ for almost all $x \in X$; that is, a mapping $f: X \to R$ is in $R^{(X)}$ if and only if there exists a finite subset F of X with $f(x) = 0$ for every $x \in X \setminus F$. Then $R^{(X)}$ is an abelian group with respect to the operation $+$ defined by

$$(f + g)(x) = f(x) + g(x)$$

for every $f, g \in R_R^{(X)}$ and every $x \in X$. The abelian group $R^{(X)}$ becomes a free right R-module $R_R^{(X)}$ with respect to the right scalar multiplication defined by

$$(fr)(x) = f(x)r$$

for every $f, g \in R_R^{(X)}$, $x \in X$ and $r \in R$.

For every fixed $x_0 \in X$, let $\delta_{x_0}: X \to R$ be the mapping defined by

$$x \mapsto \begin{cases} 1_R & \text{if } x = x_0, \\ 0_R & \text{if } x \neq x_0. \end{cases}$$

Let us prove that $\Delta := \{\delta_{x_0} \mid x_0 \in X\}$ is free set of generators for $R_R^{(X)}$. Every $f \in R_R^{(X)}$ is linear combination of finitely many δ_{x_0}'s in a unique way, because $f \in R_R^{(X)}$ implies that there exists a finite subset F of X with $f(x) = 0$ for every $x \in X \setminus F$, and then

$$f = \sum_{x_0 \in F} \delta_{x_0} f(x_0).$$

We leave to the reader the proof that this is the unique way of expressing f as a linear combination of the δ_x's. Thus $R_R^{(X)}$ is a free right module with free set of generators Δ. Equivalently,

$$R_R^{(X)} = \bigoplus_{x_0 \in X} \delta_{x_0} R$$

is a direct sum of its cyclic submodules $\delta_{x_0} R$, where, for every $x_0 \in X$, $R_R \cong \delta_{x_0} R$ as right R-modules via the isomorphism

$$R_R \to \delta_{x_0} R$$
$$r \mapsto \delta_{x_0} r.$$

Thus $R_R^{(X)}$ is isomorphic to the direct sum of $|X|$ copies of the module R_R.

We have the same also on the left. The abelian group $R^{(X)}$ is a free left R-module $_R R^{(X)}$ with respect to the left scalar multiplication defined by

$$(rf)(x) = r(f(x))$$

for every $f, g \in R_R^{(X)}$, $x \in X$ and $r \in R$. Also in this case, the set Δ is a free set of generators.

2.1 Free Modules

Example 2.1.4
Let R be a ring. Then R_R is a free right R-module with free set of generators $\{1_R\}$.

> **Proposition 2.1.5** (Universal Property of Free Modules)
> Let M_R be a free right R-module, X a free set of generators for M_R and $\varepsilon \colon X \to M_R$ the embedding of X into M_R. Then, for every right R-module M'_R and every mapping $f \colon X \to M'_R$, there exists a unique right R-module homomorphism $\tilde{f} \colon M_R \to M'_R$ making the diagram
>
> $$\begin{array}{ccc} X & \xrightarrow{f} & M'_R \\ {\scriptstyle \varepsilon}\downarrow & \nearrow {\scriptstyle \tilde{f}} & \\ M_R & & \end{array}$$
>
> commute, that is, such that $\tilde{f} \circ \varepsilon = f$.

Proof We must prove that a $\tilde{f}$ with this property exists and is unique.
Existence. If $m \in M_R$, then m can be written in the form $m = \sum_{x \in X} x \cdot r_x$, with $r_x \in R$ almost all zero, in a unique way. Define

$$\tilde{f} \colon M_R \to M'_R$$
$$m = \sum_{x \in X} x \cdot r_x \mapsto \sum_{x \in X} f(x) \cdot r_x$$

Then, for every $x \in X$,

$$\tilde{f}(\varepsilon(x)) = \tilde{f}(x) = \tilde{f}(x \cdot 1_R) = f(x) \cdot 1_R = f(x),$$

so that the diagram commutes. We need to prove that $\tilde{f}$ is a right R-module homomorphism. If $m_1, m_2 \in M_R$, then

$$m_1 = \sum_{x \in X} x \cdot r_x^1 \quad \text{and} \quad m_2 = \sum_{x \in X} x \cdot r_x^2,$$

so that

$$m_1 + m_2 = \sum_{x \in X} x \cdot (r_x^1 + r_x^2)$$

and

$$\tilde{f}(m_1 + m_2) = \sum_{x \in X} f(x) \cdot (r_x^1 + r_x^2) =$$
$$= \sum_{x \in X} f(x) \cdot r_x^1 + \sum_{x \in X} f(x) \cdot r_x^2 = \tilde{f}(m_1) + \tilde{f}(m_2)$$

and

$$\tilde{f}(m_1 \cdot r) = \tilde{f}\left(\left(\sum_{x \in X} x \cdot r_x^1\right) \cdot r\right) = \tilde{f}\left(\sum_{x \in X} (x \cdot r_x^1) \cdot r\right) =$$
$$= \tilde{f}\left(\sum_{x \in X} x \cdot (r_x^1 r)\right) = \sum_{x \in X} f(x) \cdot (r_x^1 r) =$$
$$= \sum_{x \in X} (f(x) \cdot r_x^1) \cdot r = \left(\sum_{x \in X} f(x) \cdot r_x^1\right) \cdot r = \tilde{f}(m_1) \cdot r.$$

Uniqueness. Let $g \colon M_R \to M'_R$ be any right module homomorphism such that $g \circ \varepsilon = f$. We must show that $g = \tilde{f}$. If $m \in M_R$, then $m = \sum_{x \in X} x \cdot r_x$, with $r_x \in R$ almost all zero, in a unique way, and

$$g(m) = g\left(\sum_{x \in X} x \cdot r_x\right) = g\left(\sum_{x \in X} \varepsilon(x) \cdot r_x\right) = \sum_{x \in X} (g \circ \varepsilon(x)) \cdot r_x =$$
$$= \sum_{x \in X} f(x) \cdot r_x = \tilde{f}(m).$$

Thus $g = \tilde{f}$.

□

Corollary 2.1.6
If M_R is a free right module with free set X of generators, then $M_R \cong R_R^{(X)}$.

Proof The right R-module $R_R^{(X)}$ is a free module with free set of generators $\Delta = \{\delta_x \colon x \in X\}$. Let

$$f \colon X \to R_R^{(X)}$$
$$x \to \delta_x$$

2.1 Free Modules

By the universal property of free modules, there exists a unique morphism $\varphi \colon M_R \to R_R^{(x)}$ such that the following diagram commutes

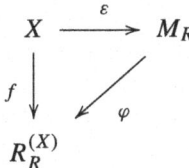

Consider now

$$g \colon \Delta \to M_R$$
$$\delta_x \mapsto x$$

By the universal property of free modules, there exists $\psi \colon R_R^{(X)} \to M_R$ such that the following diagram is commutative

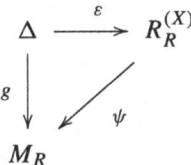

Consider now the composite mapping $\psi \circ \varphi \colon M_R \to M_R$. The mapping $\psi \circ \varphi$ sends any element $x \in X$ to x. Therefore $\psi \circ \varphi$ and 1_{M_R} are two homomorphisms of M_R into M_R such that $\psi \circ \varphi \circ \varepsilon = \varepsilon$ and $1_{M_R} \circ \varepsilon = \varepsilon$. Hence, by the uniqueness in the Universal Property of free modules, we get that $\psi \circ \varphi = 1_{M_R}$. Similarly, $\varphi \circ \psi = 1_{R_R^{(X)}}$. □

Notice that the zero module is free with free set of generators $\emptyset$. If $\varepsilon \colon \emptyset \to \{0\}$ is the unique mapping from the empty set to a zero module, then ε has the universal property of free modules. Moreover $R^{(\emptyset)} \subseteq R^{\emptyset} = \{f \colon \emptyset \to R\} = \{0\}$.

When R is a division ring, every right R-module, that is, every right vector space over the division ring R, is free. For this, we need a:

Crash Course of Linear Algebra Over Non-commutative Division Rings Let us briefly recall some elementary notions of linear algebra. The reader is definitely an expert on the elementary theory of vector spaces over a field k: vector spaces over k (they are exactly what we have called k-modules), linear transformations (they are exactly what we have called k-module morphisms), the concept of set of generators, linear combinations, linear independence and bases. The readers know that any two bases of a vector space over k have the same cardinality, and that this cardinality is called the dimension of the vector space. We know that if we have a linear transformation f between two vector spaces V and W over k of finite dimensions n and m respectively and we fix an ordered basis for V and an ordered basis for W, we can associate to f a $m \times n$ matrix with entries in k. We know the rank of a linear transformation f (it is the dimension of the image of f), bilinear mappings, the determinant of a square matrix, its minimal polynomial, the characteristic polynomial, eigenvectors and eigenvalues and so on. Assume now that k is not a field, but a division ring, and consider right vector spaces over k, that is, right k-modules. It is very easy to see that all the previous concepts hold for right vector spaces over a division ring, until when bilinear mappings enter the picture. Bilinearity is a notion concerning modules over commutative rings, because we have that $\beta(\lambda v, \mu w)$ must be equal to both $\lambda \beta(v, \mu w) = \mu \lambda \beta(v, w)$ and $\mu \beta(\lambda v, w) = \lambda \mu \beta(v, w)$.

Thus, over an arbitrary division ring k we still have linear transformations (they are the right k-module morphisms), sets of generators (again, we have already defined them for modules over arbitrary rings), linear combinations (that is, expressions of the form $\sum_{i=1}^{n} v_i \lambda_i$, where the v_i's belong to a right vector space V_k and the λ_i's are in the division ring k), linear independence (a subset X of V_k is linearly independent if and only if it is a free set of generators for the subspace of V it generates), bases (i.e., free sets of generators). Any two bases of a right vector space over a division ring k have the same cardinality (same proof as in the case of a commutative k), and this cardinality is called the dimension of the right vector space. If we have a linear transformation f between two right vector spaces V_k and W_k, $\{v_1, \ldots, v_n\}$ is a basis of V_k and $\{w_1, \ldots, w_m\}$ is a basis of W_k, we can associate to f the $m \times n$ matrix $A_f = (\lambda_{i,j})_{i,j}$, where $f(v_j) = \sum_{i=1}^{n} w_i \lambda_{i,j}$. Also in this case if $v = \sum_{j=1}^{n} v_j a_j$ is an arbitrary element of V_k and $\begin{pmatrix} a_1 \\ \vdots \\ a_n \end{pmatrix}$ is the $n \times 1$ matrix whose entries are the coefficients of v as a linear combination of $v_1, \ldots, v_n$, then the $m \times 1$ matrix $A_f \begin{pmatrix} a_1 \\ \vdots \\ a_n \end{pmatrix}$ is the matrix whose entries are the coefficients of $f(v)$ as a linear combination of $w_1, \ldots, w_m$. Notice that if $f \colon V_k \to W_k$ and $g \colon W_k \to Y_k$, then $A_{g \circ f} = A_g A_f$.

It is interesting to stress what happens for *left* vector spaces in this case. If we have a linear transformation f between two left vector spaces $_kV$ and $_kW$, it is more convenient to write f on the right, so that the image of $v \in {}_kV$ is $(v)f \in W$. Assume that $_kV$ and $_kW$ are finite-dimensional, that $\{v_1, \ldots, v_n\}$ is a basis of V_k and $\{w_1, \ldots, w_m\}$ is a basis of W_k. Then we can associate to f the $n \times m$ matrix $A_f = (\lambda_{i,j})_{i,j}$, where $(v_i)f = \sum_{j=1}^{m} \lambda_{i,j} w_j$. Also in this case if $v = \sum_{i=1}^{n} a_i v_i$ is an arbitrary element of $_kV$ and $(a_1 \ \ldots \ a_n)$ is the $1 \times n$ matrix whose entries are the coefficients of v as a linear combination of $v_1, \ldots, v_n$, then the $1 \times m$ matrix $(a_1 \ \ldots \ a_n) A_f$ is the matrix whose entries are the coefficients of $(v)f$ as a linear

2.1 Free Modules

combination of $w_1, \ldots, w_m$. If $f\colon {}_kV \to {}_kW$ and $g\colon {}_kW \to {}_kY$, then $A_{f \circ g} = A_f A_g$.

The rank of a linear transformation f is the dimension of the image of f also when the division ring k is non-commutative. The difficulties in the non-commutative case appear when bilinear mappings and determinant, which are multilinear mappings, are introduced. There are notions of right determinant and left determinant, due to Dieudonné, one is linear on the columns and the other on the rows, but they are not easy to handle. Consequently, it becomes impossible to deal (at least easily) with the minimal polynomial, the characteristic polynomial, eigenvectors and eigenvalues. But, until the appearance of bilinear mappings and determinant, the passage from the commutative case to the non-commutative one is very smooth.

As we have already said, every module over a division ring, that is, every right vector space and every left vector space, is free.

Exercise 2.1.7
Let R be a ring. Prove that every right R-module is free if and only if $R = 0$ or R is a division ring.

[*Hint:* The statement is trivial if $R = 0$. Hence let R be a non-zero ring, so that there exists a maximal right ideal I of R. Then $M_R := R/I$ is a *simple* right R-module, that is, a module M_R with exactly two submodules: the improper submodule M_R and the zero submodule. If every right R-module is free, the module M_R is free and finitely generated. Thus $R/I \cong R_R^n$ for some natural number n. Since $R/I \neq 0$, it follows that $n > 0$. If $n > 1$, then R^2 has more than two submodules. Thus $n = 1$, so $R/I \cong R_R$. Thus R has no nontrivial right ideals, so that R is a division ring.]

Let us go back to the study of free modules over arbitrary rings. Recall that $|X|$ denotes the cardinality of a set X.

> **Corollary 2.1.8**
> If M_R and N_R are free right R-modules with free sets of generators X and Y, respectively, and $|X| = |Y|$, then $M_R \cong N_R$.

Proof By Corollary 2.1.6, both M_R and N_R are isomorphic to $R_R^{(X)}$. □

If M_R is a free module, the cardinality of any free set of generators of M_R is called a *rank* of the free module M_R.

> **Proposition 2.1.9**
> Let M_R be a free right R-module. If M_R is finitely generated, then every free set of generators of M_R is finite.

Proof Let M_R be free and finitely generated. Let X be any free set of generators of M_R. Since M_R is finitely generated, there is a finite set $Y \subseteq M_R$ that generates M_R, that is, there is a finite subset $Y = \{y_1, \ldots, y_n\}$ of M_R with $YR = M_R$. Then for every $i = 1, \ldots, n$ there is a finite subset F_i of X such that y_i is a linear combination of the elements of F_i. Set $F := \cup_{i=1}^n F_i$. Then every y_i is a linear combination of elements in F, so that $M_R = YR \subseteq FR = M_R$. Hence F is a set of generators of M_R. Let us prove that the subset F of X is equal to X. If $x \in X$, then x can be written both as the linear combination $x \cdot 1_R$ and as a linear combination of elements of F. Since every element of M_R can be written in a unique way as a linear combination of elements of X, it follows that $x \in F$. This proves that $F = X$, so that X is finite. □

Proposition 2.1.10
Let R be a ring and let M_R be a right R-module with an infinite set I of generators. If no proper subset of I generates M_R, then every set of generators of M_R has cardinality $\geq |I|$.

Proof Let J be another set of generators of the module M_R. We must show that $|J| \geq |I|$. For every $a \in J$ there is a finite subset I_a of I such that a can be written as a linear combination of elements of I_a. Thus $\bigcup_{a \in J} I_a$ is a subset of I that generates M_R. By hypothesis, $\bigcup_{a \in J} I_a = I$. If J is finite, then $\bigcup_{a \in J} I_a = I$ is finite, a contradiction. Therefore J must be infinite, so that

$$|I| = |\bigcup_{a \in J} I_a| \leq \aleph_0 |J| = |J|.$$

□

Corollary 2.1.11
Let R be a ring and let M_R be a free right R-module. If X is an infinite free set of generators of M_R, then every free set of generators of M_R has cardinality $|X|$.

Hence, the rank of a free module with an infinite free set of generators is uniquely defined (Corollary 2.1.11), while the only thing we can say about a finitely generated free module is that every free set of generators is finite (Proposition 2.1.9).

2.2 Further Properties of Free Modules

We now abandon the study of the uniqueness of the rank for free modules for a while, because we must introduce some further elementary property of free modules. We say that a *module M_R is a homomorphic image of a module N_R* if there is an epimorphism $N_R \to M_R$, that is, if $M_R \cong N_R/K$ for some submodule K of N_R.

> **Proposition 2.2.1**
> *Let R be a ring. Then:*
>
> (i) *Every right R-module is a homomorphic image of a free right R-module.*
> (ii) *Every finitely generated right R-module is a homomorphic image of a free finitely generated right R-module.*

Proof Let M_R be a right R-module and let $X \subseteq M_R$ be a set of generators. Consider the inclusion mapping $\varepsilon \colon X \to M_R$ and apply the universal property of free modules to the free module $R_R^{(X)}$ with free set Δ_X of generators (Proposition 2.1.5). Thus there is a unique homomorphism $f \colon R_R^{(X)} \to M_R$ such that $f(\delta_x) = x$ for all $x \in X$. The image of f is a submodule of M_R that contains X, hence f is an epimorphism.

Now, to prove (i), let M_R be an arbitrary right R-module and $X = M_R$. Then $f \colon R_R^{(X)} \to M_R$ is an epimorphism. To prove (ii), let X be a finite set of generators for M_R. Then $f \colon R_R^{(X)} \to M_R$ is onto and $R_R^{(X)}$ is free and finitely generated. □

In the next example we fix another notation for the module $R_R^{(X)}$ in the case in which X is a finite set of cardinality n.

Example 2.2.2
Let R be a ring and let n be a positive integer. Let R^n denote the set of all n-tuples of elements of R. Then R_R^n is a free right R-module of rank n, and $_RR^n$ is a free left R-module of rank n. Observe that the two R-modules R_R^n and $_RR^n$ have the same addition, but in the first we have a right scalar multiplication defined as $(r_1, \ldots, r_n)r = (r_1 r, \ldots, r_n r)$ for every $(r_1, \ldots, r_n) \in R^n$ and every $r \in R$; in the second the left scalar multiplication is given by $r(r_1, \ldots, r_n) = (rr_1, \ldots, rr_n)$.

▶ **Remark 2.2.3** The idea of Sect. 1.8, that every morphism between finite direct sums of modules M_i and N_j can be described as a matrix of morphisms $M_i \to N_j$, can be specialized to the case in which all the modules M_i and N_j are isomorphic to R_R. Notice that $\mathrm{Hom}(R_R, R_R) \cong R$, because every morphism $f \colon R_R \to R_R$ is given by left multiplication by an element $a_f \in R$, namely for every homomorphism $f \colon R_R \to R_R$ there exists an element $a_f \in R$ such that $f(r) = a_f r$ for every $r \in R_R$. Thus a morphism $R_R^n \to R_R^m$ is completely described by an $m \times n$ matrix.

More precisely, let m and n be two positive integers. We have an abelian group isomorphism $\mathrm{Hom}(R_R^n, R_R^m) \cong \mathbb{M}_{m \times n}(R)$, where $\mathbb{M}_{m \times n}(R)$ is the additive abelian group of all $m \times n$ matrices with entries in R. If we denote the elements of R_R^n and R_R^m as columns, that is, as $n \times 1$ and $m \times 1$ matrices respectively, every R-linear mapping $f \colon R_R^n \to R_R^m$ is given by left multiplication by a matrix $A_f \in \mathbb{M}_{m \times n}(R)$ uniquely determined by f:

$$f \colon \begin{pmatrix} r_1 \\ \vdots \\ r_n \end{pmatrix} \mapsto A_f \begin{pmatrix} r_1 \\ \vdots \\ r_n \end{pmatrix}$$

for every $r_1, \ldots, r_n \in R$.

In particular, if $n = m$, then

$$\mathrm{Hom}(R_R^n, R_R^n) = \mathrm{End}(R_R^n) \cong \mathbb{M}_{n \times n}(R)$$

(the isomorphism is a ring isomorphism). In this case, for every $f, g \in \mathrm{End}(R_R^n)$, we have $A_{fg} = A_f A_g$, and R^n becomes an $\mathbb{M}_{n \times n}(R)$-$R$-bimodule ${}_{\mathbb{M}_{n \times n}(R)} R^n{}_R$. Here, in the formula $A_{fg} = A_f A_g$, we write the composite mapping as fg because we write morphisms on the left for right R-modules.

The situation is similar for left R-modules: for any two positive integers n and m, we have an abelian group isomorphism

$$\mathrm{Hom}({}_R R^n, {}_R R^m) \cong \mathbb{M}_{n \times m}(R),$$

because every R-linear mapping $f \colon {}_R R^n \to {}_R R^m$ is given by right multiplication by a matrix $A_f \in \mathbb{M}_{n \times m}(R)$ uniquely determined by f:

$$f \colon (r_1 \ \ldots \ r_n) \mapsto (r_1 \ \ldots \ r_n) A_f$$

for every $r_1, \ldots, r_n \in R$, (notice that here we write the elements of ${}_R R^n$ as rows and the matrix A_f on the right).

If $n = m$, then

$$\mathrm{Hom}({}_R R^n, {}_R R^n) = \mathrm{End}({}_R R^n) \cong \mathbb{M}_{n \times n}(R),$$

and this isomorphism is a ring isomorphism. In this case, for every $f, g \in \mathrm{End}({}_R R^n)$ we have $A_{fg} = A_f A_g$ (where we write morphisms on the right), and R^n becomes an R-$\mathbb{M}_{n \times n}(R)$-bimodule ${}_R R^n{}_{\mathbb{M}_{n \times n}(R)}$.

Other Free Algebraic Structures

Free Monoids

Let M be a monoid. The intersection of an arbitrary family of submonoids of M is also a submonoid of M. If $X \subseteq M$, the intersection of all submonoids of M containing X is the smallest submonoid of M containing X and is called the submonoid of M generated by X. We will denote it by $\langle X \rangle$.

It is easy to verify that $\langle X \rangle = \left\{ 1_M, \prod_{i=1}^n x_i \mid n \geq 1, \ x_1, \ldots, x_n \in X \right\}$.

We say that $X \subseteq M$ is a *set of generators* of M if $M = \langle X \rangle$, and say that X is a *free set of generators* of M if X generates M and every element of M can be written as a product of elements of X in a unique way. (The identity 1_M can be viewed as the product of zero elements of X, for every subset X of M_R.)

Example 2.2.4
Let $X \neq \emptyset$ be a set. A *word of length* $n \geq 0$ *in the alphabet* X is any n-tuple $(x_1 \cdots x_n)$ of elements of X; we will denote $(x_1 \cdots x_n)$ by $x_1 \ldots x_n$. There is only one 0-tuple, namely the empty word $\emptyset$. Set $W := \bigcup_{n \geq 0} W_n$, where W_n is the set of all words of length n in the alphabet X (i.e., $W_n = X^n$). Thus W is the disjoint union of the W_n's.

Now define a multiplication (called *concatenation*) between words of length n and m respectively as follows: if $n, m \geq 0$, $x_1 \ldots x_n \in W_n$ and $y_1 \ldots y_m \in W_M$, set $x_1 \ldots x_n \cdot y_1 \ldots y_m := x_1 \ldots x_n y_1 \ldots y_m$. The empty word $\emptyset$ is the two-sided identity with respect to concatenation, so that W becomes a monoid, called *the monoid of the words in the alphabet* X. It is easy to check that W is a free monoid with free set of generators W_1.

We have a universal property for free monoids as well.

Theorem 2.2.5 (Universal Property of Free Monoids)
Let F be a free monoid with free set of generators $X \subseteq F$ and let $\varepsilon \colon X \to F$ be the inclusion map. Then, for every monoid M and every mapping $f \colon X \to M$, there exists a unique monoid homomorphism $g \colon F \to M$ such that $g \circ \varepsilon = f$.

From this universal property, we obtain that every free monoid with free set X of generators is isomorphic to the monoid W_X of all words in the alphabet X. Every monoid is a homomorphic image of a free monoid.

Exercise 2.2.6
Let $I_{\mathbf{Set}} \colon \mathbf{Set} \to \mathbf{Set}$ be the identity functor.
Let

$$M \colon \mathbf{Set} \to \mathbf{Mon}$$
$$X \mapsto W_X$$
$$(f \colon X \to Y) \mapsto (\widetilde{f} \colon W_X \to W_Y),$$

be the covariant functor that associates to each set X the free monoid on X, and to each mapping f the mapping $\widetilde{f}$ given by the universal property.

Let $F\colon \mathbf{Mon} \to \mathbf{Set}$ be the forgetful functor, so that the composition

$$F \circ M \colon \mathbf{Set} \to \mathbf{Set}$$

is a covariant functor.

Show that there is a natural transformation $\varepsilon\colon I_{\mathbf{Set}} \to F \circ M$ given by the canonical embedding

$$\varepsilon_X \colon X \hookrightarrow W_X.$$

Free Groups

We now pass to consider the *free group* on a set $X \neq \emptyset$. It will be the analogue of free monoid but in the category of groups instead of the category of monoids. Fix a set $X \neq \emptyset$. Let X^* be another set disjoint from X and with the same cardinality as X. There is a bijection $f\colon X \to X^*$, and we will denote by x^* the image of any element $x \in X$ via f.

Let W be the free monoid on the set $X \cup X^*$. Let $\sim$ be the congruence on the monoid W generated by the $2|X|$ relations $xx^* \sim 1_W$ and $x^*x \sim 1_W$ for every $x \in X$. The free group on the set X will be $G := W/\sim$.

Let us show that G is a group. It is a monoid because it is a monoid modulo a congruence. Its identity is $1_G = 1_{W/\sim} = [1_W]_\sim$. An arbitrary element of G is of the form $[y_1 \ldots y_n]_\sim$ with $y_1, \ldots, y_n \in X \cup X^*$. If y_i is an element $x \in X$, set $y_i^* := x^*$; if y_i is an element $x^* \in X^*$, set $y_i^* := x$. It is easily checked that the inverse of $[y_1 \ldots y_n]_\sim$ is $[y_n^* \ldots y_1^*]_\sim$. Thus G is a group.

A universal property holds in this case also. Let us prove that if H is any group and $f\colon X \to H$ is any mapping, then there exists a group morphism of G into H that extends f. (We leave the proof of the uniqueness of this group morphism to the reader.)

Given a mapping $f\colon X \to H$, extend f to $f_0\colon X \cup X^* \to H$ defining $f_0(x^*) = (f(x))^{-1}$ for every $x \in X$. By the universal property of free monoids, f_0 extends uniquely to a monoid homomorphism $\widetilde{f_0}\colon W \to H$.

Let us show that $\widetilde{f_0}$ induces a monoid homomorphism $W/\sim\ \to H$. We must prove that if $w_1, w_2 \in W$ and $w_1 \sim w_2$, then $\widetilde{f_0}(w_1) = \widetilde{f_0}(w_2)$. Since $\sim$ is the congruence generated by the $2|X|$ relations $xx^* \sim 1_W$ and $x^*x \sim 1_W$, it suffices to show that $\widetilde{f_0}(xx^*) = \widetilde{f_0}(1_W)$ and $\widetilde{f_0}(x^*x) = \widetilde{f_0}(1_W)$. Now $\widetilde{f_0}(xx^*) = \widetilde{f_0}(x)\widetilde{f_0}(x^*) = f_0(x)f_0(x^*) = f(x)(f(x))^{-1} = 1_H = \widetilde{f_0}(1_W)$. Similarly $\widetilde{f_0}(x^*x) = \widetilde{f_0}(1_W)$. Thus $\widetilde{f_0}$ induces a monoid homomorphism $W/\sim\ \to H$, which is the required extension of the mapping $f\colon X \to H$ to a group homomorphism of G into H.

Notice that if G is the free group on a set X of cardinality one, then $G \cong \mathbb{Z}$.

Free k-Algebras

Let k be a commutative ring. We now introduce k-algebras, which also can be defined in two equivalent ways.

2.2 Further Properties of Free Modules

Definition 2.2.7

A *k-algebra* is a pair (R, ϕ), where R is a ring and $\phi: k \to R$ is a ring homomorphism whose image is contained in the center $Z(R)$ of R.
A *k-algebra homomorphism* $f: (R, \phi) \to (S, \psi)$ is a ring homomorphism $f: R \to S$ that makes the diagram

$$\begin{array}{ccc} & k & \\ \phi \swarrow & & \searrow \psi \\ R & \xrightarrow{f} & S \end{array}$$

commute.

Definition 2.2.8

A *k-algebra* $(R, +, \cdot, *)$ is a ring $(R, +, \cdot)$, which is also a *k*-module $R_k = (R, +, *)$, and in which $(r \cdot s) * \lambda = (r * \lambda) \cdot s = r \cdot (s * \lambda)$ for every $r, s \in R$ and $\lambda \in k$. A *k-algebra homomorphism* $f: R \to S$ is a mapping that is both a ring homomorphism and a *k*-module homomorphism.

Given a *k*-algebra R and a subset X of R, the intersection of all *k*-subalgebras of R (i.e., subrings of R that are also submodules of R_k) that contain X is the *k*-subalgebra of R generated by X. Denote it by $k\langle X \rangle$. The algebra $k\langle X \rangle$ turns out to be the set of all linear combinations with coefficients in k of finite products of elements of X, that is,

$$k\langle X \rangle = \{ \sum_{\text{finite}} x_{i_1} \ldots x_{i_n} \lambda_{i_1 \ldots i_n} \mid x_{i_t} \in X, \lambda_{i_1 \ldots i_n} \in k \}.$$

Hence $X \subseteq R$ is a *set of generators* of the *k*-algebra R if and only if every element $r \in R$ can be written as a finite sum of products of the form $x_1 \ldots x_n \lambda$ with $x_1 \ldots x_n \in X$ and $\lambda \in k$. The set of generators X is *free* if this can be done in a unique way.

Example 2.2.9

Take a commutative ring k, and a set $X \neq \emptyset$. Construct the free monoid W_X. Construct the free *k*-module on the set W_X:

$$k_k^{(W_X)} = \left\{ \sum_{i=1}^{n} w_i \lambda_i \mid n \geq 1, \ w_i \text{ word in the alphabet } X, \ \lambda_i \in k \right\}.$$

Then $k_k^{(W_X)}$ is the ring of polynomials with coefficients in k in the set of non-commuting indeterminates X, and is a free *k*-algebra with free set of generators X. It is usually denoted by $k\langle X \rangle$.

Example 2.2.10
If $X = \{x, y\}$, then $k\langle X\rangle = k\langle x, y\rangle$ turns out to be the set of all polynomials $1 \cdot \lambda_0 + x\lambda_1 + y\lambda_2 + x^2\lambda_3 + xy\lambda_4 + yx\lambda_5 + y^2\lambda_6 + \ldots$. For any degree n, $k\langle x, y\rangle$ has 2^n monic monomials.

Exercise 2.2.11
Determine the free commutative k-algebras. Here k is a commutative ring, and "commutative k-algebra R" means "ring R that is commutative and is a k-algebra". Equivalently, a commutative k-algebra can be defined as a pair (R, φ), where $\varphi \colon k \to R$ is a ring homomorphism and R is a commutative ring.

2.3 Further Terminology About Categories

Let $F \colon \mathcal{C} \to \mathcal{D}$ be a covariant (contravariant) functor. For all $C, C' \in \mathrm{Ob}\,\mathcal{C}$, the functor F induces a mapping

$$F_{CC'} \colon \mathrm{Hom}_{\mathcal{C}}(C, C') \to \mathrm{Hom}_{\mathcal{D}}(F(C), F(C'))$$

($F_{CC'} \colon \mathrm{Hom}_{\mathcal{C}}(C, C') \to \mathrm{Hom}_{\mathcal{D}}(F(C'), F(C))$), defined by $F_{CC'}(f) = F(f)$ for every $f \colon C \to C'$. If all these mappings are injective, F is called a *faithful functor*. If all these mappings are onto, F is called a *full functor*. If all these mappings are bijections, the functor F is called a *fully faithful functor*. A functor $F \colon \mathcal{C} \to \mathcal{D}$ is *essentially surjective* (or *dense*) if for every $D \in \mathrm{Ob}\,\mathcal{D}$ there exists $C \in \mathrm{Ob}\,\mathcal{C}$ such that $F(C) \cong D$.

Examples 2.3.1
(1) The forgetful functor $F \colon \mathbf{Grp} \to \mathbf{Set}$ is faithful but not full.
(2) Let $W \colon \mathbf{Set} \to \mathbf{Mon}$ be the functor that associates to each set X the monoid W_X of words in the alphabet X. It is easily seen that W is faithful. It is not full, because if $X = X'$ has a unique element a, then $\mathrm{Hom}_{\mathbf{Mon}}(W_{\{a\}}, W_{\{a\}})$ has at least two elements (the identity $\iota_{W_{\{a\}}}$ and the trivial morphism that sends every element of $W_{\{a\}}$ to $1_{W_{\{a\}}}$) and $\mathrm{Hom}_{\mathbf{Set}}(\{a\}, \{a\})$ has one element, so that the mapping $W_{\{a\}\{a\}}$ cannot be an onto mapping.

It is possible to prove the following Lemma:

Lemma 2.3.2
The following conditions are equivalent for a covariant functor $F \colon \mathcal{C} \to \mathcal{D}$:

(i) *There exists a covariant functor $G \colon \mathcal{D} \to \mathcal{C}$ such that $G \circ F$ is naturally isomorphic to the identity functor $I_{\mathcal{C}}$ and $F \circ G$ is naturally isomorphic to the identity functor $I_{\mathcal{D}}$.*
(ii) *The functor F is full, faithful and essentially surjective.*

2.3 Further Terminology About Categories

A functor F satisfying the equivalent conditions of this Lemma is called a *category equivalence* (or simply an *equivalence*). Two categories $\mathcal{C}$ and $\mathcal{D}$ are said to be *equivalent* if there exists an equivalence $F: \mathcal{C} \to \mathcal{D}$. Clearly, isomorphic categories are equivalent.

Let $\mathcal{C}^{\mathrm{op}}$ denote the opposite category (that is, the dual category) of $\mathcal{C}$ (Example 1.2.2(6)). An equivalence between the categories $\mathcal{C}^{\mathrm{op}}$ and $\mathcal{D}$ is called a *duality* between $\mathcal{C}$ and $\mathcal{D}$. More precisely, a *duality* is a contravariant functor $F: \mathcal{C} \to \mathcal{D}$ satisfying the following equivalent conditions: (a) There exists a contravariant functor $G: \mathcal{D} \to \mathcal{C}$ such that $G \circ F$ is naturally isomorphic to the identity functor $I_\mathcal{C}$ and $F \circ G$ is naturally isomorphic to the identity functor $I_\mathcal{D}$; (b) The functor F is full, faithful, and for every $D \in \mathrm{Ob}\,\mathcal{D}$ there exists $C \in \mathrm{Ob}\,\mathcal{C}$ such that $F(C) \cong D$.

Exercise 2.3.3
Let k be a division ring, **Vect**-k (k-**Vect**) the category of all right (left, respectively) vector spaces over k and **vect**-k (k-**vect**) the category of all right (left) vector spaces over k that are of finite dimension.
(1) Show that the contravariant functor

$$D := \mathrm{Hom}(-, {}_k k_k): \mathbf{Vect}\text{-}k \to k\text{-}\mathbf{Vect}$$

is not a duality.
(2) Show that the restriction $D: \mathbf{vect}\text{-}k \to k\text{-}\mathbf{vect}$ is a duality, but not an isomorphism of categories between $(k\text{-}\mathbf{vect})^{\mathrm{op}}$ and $\mathbf{vect}\text{-}k$.
[*Hint for (1)*: If $\dim V_k = \aleph \geq \aleph_0$, then $|\mathrm{Hom}(V_k, {}_k k_k)| = |k^\aleph| = |k|^\aleph \geq 2^\aleph$, while if ${}_k W$ has dimension $\aleph_0$, then

$$|{}_k W| \leq |\bigcup_{n \geq 0} k^n| \leq \sum_{n \geq 0} |k^n| \leq \aleph_0 |k|.$$

Hence $|{}_k W| = \aleph_0$ for $|k| \leq \aleph_0$.]

Exercise 2.3.4
Let **Mon** be the category of monoids (Exercise 1.7.5). Let $\mathbf{Cat}_1$ be the category whose objects are all the categories with exactly one object, i.e., all the categories $\mathcal{C}$ with $|\mathrm{Ob}\,\mathcal{C}| = 1$, and with set of morphisms $F: \mathcal{C} \to \mathcal{D}$ between any two such categories $\mathcal{C}$ and $\mathcal{D}$ the set of all covariant functors $F: \mathcal{C} \to \mathcal{D}$. Define a category equivalence between **Mon** and $\mathbf{Cat}_1$.

Exercise 2.3.5
Let $f: R \to S$ be a ring homomorphism.
(1) Define a canonical functor $F: \mathrm{Mod}\text{-}S \to \mathrm{Mod}\text{-}R$ associated to f.
(2) Show that the functor F defined in (1) is faithful.
For any S-module M, the R-module $F(M)$ is said to be obtained from M by *restriction of scalars*.

(The name "restriction of scalar" comes from the case where $R \subseteq S$ and $f: R \to S$ is the inclusion. In this cases, in passing from the S-module structure to the R-module structure, we really restrict the scalars from S to R.)

Exercise 2.3.6
Let **Ring** be the category of rings with identity, **Rng** be the category of rings not-necessarily with an identity, and **Ring**$_a$ be the category of *rings with augmentation*, whose objects are the morphisms $\pi: R \to \mathbb{Z}$ in **Ring** with R any ring with identity and whose morphisms $f: \pi \to \pi'$ are the morphisms $f: R \to R'$ of rings with identity for which $\pi' f = \pi$. Show that if we associate, like in Remark 1.1.2(2), to any "ring without identity" $(R, +, \cdot)$ the ring with identity $R \oplus \mathbb{Z}$, one gets a category equivalence between the categories **Rng** and **Ring**$_a$.

Exercise 2.3.7
Let $F, G: \mathcal{C} \to \mathcal{D}$ be two functors, $\eta: F \to G$ be a natural isomorphism, and $A: \mathcal{C}' \to \mathcal{C}$, $B: \mathcal{D} \to \mathcal{D}'$ be two further functors. Show that the functors BFA and BGA are naturally isomorphic.

Let $\mathcal{C}$ be a category. A *subcategory* $\mathcal{S}$ of $\mathcal{C}$ consists of:
(i) a subclass $\text{Ob}\,\mathcal{S}$ of $\text{Ob}\,\mathcal{C}$; and
(ii) for each pair (S, T) of objects in $\mathcal{S}$, a subset $\text{Hom}_\mathcal{S}(S, T)$ of $\text{Hom}_\mathcal{C}(S, T)$
such that
(a) $1_S \in \text{Hom}_\mathcal{S}(S, S)$ for every $S \in \text{Ob}\,\mathcal{S}$, and
(b) for every triple (S, T, U) of objects in $\mathcal{S}$ and every morphism $f \in \text{Hom}_\mathcal{S}(S, T)$ and $g \in \text{Hom}_\mathcal{S}(T, U)$, one has that $g \circ f \in \text{Hom}_\mathcal{S}(S, U)$.

These conditions ensure that a subcategory of a category is a category with respect to the composition induced by the composition of $\mathcal{C}$.

A *full subcategory* is a subcategory $\mathcal{S}$ of $\mathcal{C}$ such that $\text{Hom}_\mathcal{S}(S, T) = \text{Hom}_\mathcal{C}(S, T)$ for every pair (S, T) of objects in $\mathcal{S}$. Thus a full subcategory of a category $\mathcal{C}$ is completely determined by its class of objects. Conversely, if $\mathcal{C}$ is a category, every subclass of $\text{Ob}\,\mathcal{C}$ is the class of objects of a unique full subcategory of $\mathcal{C}$.

For example, **Ab** is a full subcategory of **Grp**.

Exercise 2.3.8
Show that the category of finite pre-ordered sets is isomorphic to the category of finite topological spaces, the full subcategory of **Top** whose objects are the topological space with only finitely many points. (If X is a finite topological space, the corresponding pre-order $\leq$ on X is defined by $x \leq y$ if and only if x belongs to the closure of the subset $\{y\}$ of X. Conversely, every closed set in a finite topological space X is a union of closures of points.)

More generally, the category of pre-ordered sets is isomorphic to the category $\mathcal{A}$ of Alexandrov topological spaces, the full subcategory of **Top** whose objects are the topological space whose topology is an Alexandrov topology. A topology is *Alexandrov* if the intersection of any family of open subsets is an open set (equivalently, if the union of any family of closed subsets is a closed subset).

2.3 Further Terminology About Categories

Exercise 2.3.9
Let $\mathcal{F}_R$ ($_R\mathcal{F}$) be the full subcategory of Mod-R (R-Mod) whose objects are all finitely generated free right (left, respectively) R-modules. Show that the functor

$$\mathrm{Hom}(-, {}_RR_R)\colon \mathcal{F}_R \to {}_R\mathcal{F}$$

is a duality. In particular, $R_R^n \cong R_R^m$ if and only if $_RR^n \cong {}_RR^m$ for every $n, m \geq 0$.

If $\mathcal{S}$ is a subcategory of $\mathcal{C}$, there is an embedding functor $E\colon \mathcal{S} \to \mathcal{C}$ that assigns to every $S \in \mathrm{Ob}\,\mathcal{S}$ the object S itself, and to every morphism f in $\mathcal{S}$ the morphism f viewed as a morphism in $\mathcal{C}$. The functor E is always faithful, for any subcategory $\mathcal{S}$ of $\mathcal{C}$. It is a full functor if and only if $\mathcal{S}$ is a full subcategory of $\mathcal{C}$.

Exercise 2.3.10
Let $\mathcal{A}$ be a category with only one object A and one morphism $f\colon A \to A$. Let **Set**$_1$ be the full subcategory of the category **Set** whose objects are all sets of cardinality 1. Show that the categories **Set**$_1$ and $\mathcal{A}$ are equivalent.

A *skeleton* of a category $\mathcal{C}$ is any full subcategory $\mathcal{S}(\mathcal{C})$ of $\mathcal{C}$ such that every object of $\mathcal{C}$ is isomorphic to exactly one object of $\mathcal{S}(\mathcal{C})$.

Exercise 2.3.11
Let $\mathcal{S}$ be a subcategory of a category $\mathcal{C}$ and let $E\colon \mathcal{S} \to \mathcal{C}$ be the embedding functor. Show that $\mathcal{S}$ is a skeleton of $\mathcal{C}$ if and only if the following two conditions hold:
(1) the functor E is a category equivalence;
(2) no two distinct objects of $\mathcal{S}$ are isomorphic.

Examples 2.3.12
(1) In the category **Set** of all sets, the full subcategory whose objects are all cardinal numbers is a skeleton for **Set**.
(2) In the category **vect**-k, where k is a field, of all finite-dimensional vector spaces over k, the subcategory whose objects are all vector spaces k^n, where n is any non-negative integer, is a skeleton for **vect**-k.
(3) Let **FinSet** be the full subcategory of **Set** whose objects are all finite sets. The full subcategory of **FinSet** whose objects are all sets $\{0, 1, 2, \ldots, n-1\}$, $n \in \mathbb{N}$, is a skeleton of **FinSet**.

A digression of Set Theory is now necessary. This topic is further developed in Appendix 1, for the interested reader.

The most popular and accepted form of axiomatic set theory is ZFC, the *Zermelo-Fraenkel set theory* with the axiom of choice. It has a single primitive ontological notion, the notion of set. That is, it treats only sets (and not classes): all individuals in the universe of discourse are sets. Sets are denoted with lower case letters. The only binary relations are equality and set membership, denoted $\in$. Thus the formula $x \in y$ indicates that x and y are sets and that x *belongs to* y (or x *is an element of* y, or x *is a member* of y). We can only use the logical symbols ($\neg, \wedge, \vee, \to, \leftrightarrow, \forall, \exists$), $=$ (equality), parentheses, lower case letters (variable symbols) and the symbol $\in$. (One must follow the rules studied in any course of mathematical logic to get well-formed formulas!)

In these notes, we need the *Von Neumann-Bernays-Gödel set theory* (NBG). NBG is a conservative extension of ZFC. The ontology of NBG includes proper classes. The members of both sets and proper classes are sets. Classes cannot be members. "Conservative extension" means that a statement in the language of ZFC is provable in NBG if and only if it is provable in ZFC, that is, any theorem in NBG which speaks only about sets is a theorem in ZFC. In NBG, quantified variables in the defining formula can range only over sets.

Let us try to be more precise. The characteristic of NBG is the distinction between proper classes and sets. NBG is a two-sorted theory, that is, two types of variables are used in NBG. Lower case letters will denote variables ranging over sets, and upper case letters will denote variables ranging over classes. The atomic sentences $a \in b$ and $a \in A$ are defined for a, b sets and A a class, but $A \in a$ or $A \in B$ are not defined for any two classes A, B. Equality can have the form $a = b$ or $A = B$. The sentence $a = A$ stands for $\forall x (x \in a \leftrightarrow x \in A)$ and is an abuse of notation. NBG can also be presented as a one-sorted theory of classes, with sets being those classes that are members of at least one other class. That is, NBG can be presented as a system having only one type of variables (class variables) with a unary relation $\mathcal{M}(A)$ ($\mathcal{M}$ stands for the German word Menge, which mean "set"), and $\mathcal{M}(A)$ indicates that A is a set. Thus $\mathcal{M}(A) \leftrightarrow \exists B(A \in B)$. Notice that NBG admits the class V of all sets, but it does not admit the class of all classes or the set of all sets. One of the axioms of NBG is the:

Axiom of Limitation of Size For any class A, there exists a set a such that $a = A$ if and only if there is no bijection between A and the class V of all sets.

This is really a powerful axiom. From this axiom, every proper class, that is, every class that is not a set, is equipotent to the class V of all sets. Moreover, the axiom of choice for classes holds, because the class of ordinals is not a set, so that there is a bijection between the ordinals and any proper class, and any class can be well-ordered (*Well-ordering Theorem*). Equivalently, if A is any class and $\sim$ is an equivalence relation on A, a class of representatives exists (This is standard form of the *Axiom of choice for classes*.) It follows that:

Every Category Has a Skeleton

Exercise 2.3.13
Let $\mathcal{S}$ be a skeleton of a category $\mathcal{C}$ and let $E: \mathcal{S} \to \mathcal{C}$ be the embedding functor.
(1) Show that the functor E has left inverses, that is, there exist functors $G: \mathcal{C} \to \mathcal{S}$ such that GE is the identity functor $I_{\mathcal{S}}$.
(2) We have already seen in Exercise 2.3.11(1) that the embedding functor E is an equivalence of categories. Show that its left inverses G also are category equivalences.

(3) Let $\mathcal{S}$ be a skeleton of a category $\mathcal{C}$ and $E \colon \mathcal{S} \to \mathcal{C}$ be the embedding functor. Similarly, let $\mathcal{S}'$ be a skeleton of another category $\mathcal{C}'$ and $E' \colon \mathcal{S}' \to \mathcal{C}'$ be the embedding. Let $F \colon \mathcal{C} \to \mathcal{C}'$ be any functor. Set $F' := G'FE \colon \mathcal{S} \to \mathcal{S}'$, where G' denotes a left inverse of E'. Show that the functors FE and $E'F'$ are naturally isomorphic. In other words, the diagram

$$\begin{array}{ccc} \mathcal{S} & \xrightarrow{F'} & \mathcal{S}' \\ E \downarrow & & \downarrow E' \\ \mathcal{C} & \xrightarrow{F} & \mathcal{C}' \end{array}$$

"is commutative up to a natural isomorphism".

(4) Let G be a left inverse of E. In the notation of (3), show that the functors F and $E'F'G$ are naturally isomorphic.
(5) Show that F is a category equivalence if and only if F' is a category isomorphism.
(6) Show that two categories $\mathcal{C}, \mathcal{C}'$ are equivalent if and only if their skeletons $\mathcal{S}, \mathcal{S}'$ are isomorphic.
(7) Show that any two skeletons of a category $\mathcal{C}$ are isomorphic categories.

Exercise 2.3.14
Prove Lemma 2.3.2.

Let $\{\mathcal{C}_\lambda \mid \lambda \in \Lambda\}$ be a family of categories indexed in a set Λ. The *product category* $\mathcal{C} = \prod_{\lambda \in \Lambda} \mathcal{C}_\lambda$ of the given family is defined as follows. The class of objects of $\mathcal{C}$ is

$$\mathrm{Ob}\,\mathcal{C} = \prod_{\lambda \in \Lambda} \mathrm{Ob}\,\mathcal{C}_\lambda = \{\,(C_\lambda)_{\lambda \in \Lambda} \mid C_\lambda \in \mathrm{Ob}\,\mathcal{C}_\lambda \text{ for all } \lambda \in \Lambda\,\}.$$

The set of morphisms between any two objects $C = (C_\lambda)_{\lambda \in \Lambda}$ and $C' = (C'_\lambda)_{\lambda \in \Lambda}$ is

$$\mathrm{Hom}_\mathcal{C}(C, C') = \prod_{\lambda \in \Lambda} \mathrm{Hom}_{\mathcal{C}_\lambda}(C_\lambda, C'_\lambda).$$

Composition is defined componentwise.

Exercise 2.3.15
Show that the categories Mod-$(\mathbb{Z}/2\mathbb{Z}) \times$ Mod-$(\mathbb{Z}/3\mathbb{Z})$ and Mod-$(\mathbb{Z}/6\mathbb{Z})$ are equivalent. More generally, show that if R and S are rings and $R \times S$ is their direct product, then Mod-$R \times$ Mod-S and Mod-$(R \times S)$ are two equivalent categories.

2.4 IBN Rings

A ring R has *IBN* (invariant basis number) if, for every $n, m \geq 0$, $R_R^n \cong R_R^m$ implies $n = m$. For instance, division rings have IBN. Trivially, the zero ring is not IBN.

Exercise 2.4.1
(1) Show that having IBN is a left/right symmetric condition, that is, a ring R has IBN if and only if $_R R^n \cong {_R R^m}$ implies $n = m$ for every $n, m \geq 0$.
(2) Notation as in Remark 2.2.3. Show that a ring R has IBN if and only if for every $n, m \geq 1$, $A \in \mathbb{M}_{n \times m}(R)$, $B \in \mathbb{M}_{m \times n}(R)$, $AB = 1_n$, $BA = 1_m$ imply $n = m$.
(3) Show that if there exists a ring homomorphism $\varphi \colon R \to S$ and the ring S has IBN, then R has IBN as well.
(4) Show that if R is a ring, I is a proper two-sided ideal in R and the quotient ring R/I has IBN, then R has IBN as well.
(5) Show that every non-zero commutative ring has IBN.

[*Hint for (3):* Apply (2). If $A \in \mathbb{M}_{n \times m}(R)$, $B \in \mathbb{M}_{m \times n}(R)$ and we apply the homomorphism φ to the entries of A and B, we get two matrices in $\mathbb{M}_{n \times m}(S)$ and $\mathbb{M}_{m \times n}(S)$ such that...]

Example 2.4.2
We will now give an example of a ring R with $R_R \cong R_R \oplus R_R$, so that in particular the ring R has not IBN. Let k be a field. Let V_k be a vector space over k of infinite dimension, so that $V_k \oplus V_k \cong V_k$. Let $R := \text{End}(V_k)$ be the endomorphism ring of V_k, so that $_R V_k$ is a R-k-bimodule. Thus there is a covariant functor $\text{Hom}(_R V_k, -) \colon \text{Mod-}k \to \text{Mod-}R$. Applying this functor to the right k-module isomorphism $V_k \oplus V_k \cong V_k$, we get a right R-module isomorphism $\text{Hom}(_R V_k, V_k) \oplus \text{Hom}(_R V_k, V_k) \cong \text{Hom}(_R V_k, V_k)$, that is, an isomorphism $R_R \oplus R_R \cong R_R$. This is an isomorphism between two free right R-modules of rank 2 and 1 respectively. Therefore R is not an IBN ring. Notice that, for this ring R, we have that $R_R \cong R_R^n$ for every $n \geq 1$. Thus $R_R^n \cong R_R^m$ for every $n, m \geq 1$.

Before stating the next example, let us recall what the two-sided ideal generated by a subset is. Let R be a ring and X a subset of R. Any intersection of two-sided ideals of R is a two-sided ideal of R. In particular, the intersection of all two-sided ideals of R containing X (there are always such ideals, for instance R itself) is a two-sided ideal containing X. It is the smallest two-sided ideal containing X, called the two-sided ideal of R *generated* by X, and denoted (X). If $X \neq \emptyset$, it is easily seen that $(X) = \{\sum_{i=1}^n r_i x_i r'_i \mid n \geq 1,\ r_i, r'_i \in R\ x_i \in X\}$. For instance, if X is the set with only one element $a \in R$, then the *principal (two-sided) ideal of R generated by a*, that is, the two-sided ideal generated by the set $X = \{a\}$, is the set of all elements of the form $\sum_{i=1}^n r_i a r'_i$, where $n \geq 1$ and $r_i, r'_i \in R$. This set can be properly larger than the set of all elements of the form rar' with $r, r' \in R$. For instance, if R is the ring of all 2×2 matrices with entries in a field k and a is the matrix $\begin{pmatrix} 1 & 0 \\ 0 & 0 \end{pmatrix}$, then R is a simple ring (Exercise 1.1.11), so that the principal ideal generated by a is not a proper ideal, but it is easily seen that the set of all elements of R of the form rar' with $r, r' \in R$ is a proper subset of R.

Example 2.4.3 (Leavitt Algebra)
Fix $m, n \geq 1$. We want to give another example of a ring R such that $R_R^n \cong R_R^m$. Let k be a field, and let

$$Z := \{x_{ij}, y_{ji} \mid i = 1, \ldots, n,\ j = 1, \ldots, m\}$$

be a set of cardinality $2mn$. Let
$$P := k \langle \{ x_{ij}, \, y_{ji} \mid i = 1, \ldots, n, \ j = 1, \ldots, m \} \rangle$$
be the free k-algebra over Z. Let $X := (x_{ij})_{i,j} \in \mathbb{M}_{n \times m}(P)$ and $Y := (y_{ji})_{j,i} \in \mathbb{M}_{m \times n}(P)$. Let I be the two-sided ideal of P generated by the $n^2 + m^2$ entries of the two matrices $XY - 1_n$ and $YX - 1_m$. Let us prove that I is a proper ideal. Consider a vector space V_k over k of infinite dimension, so that, as we have seen in Example 2.4.2, if T is its endomorphism ring $\text{End}(V_k)$, then $T_T^n \cong T_T^m$. Hence there exist two matrices $A := (t_{ij})_{i,j} \in \mathbb{M}_{n \times m}(T)$ and $B := (t'_{ji})_{j,i} \in \mathbb{M}_{m \times n}(T)$ such that $AB = 1_n$ and $BA = 1_m$. By the universal property of free k-algebras, there is a unique k-algebra homomorphism $\varphi \colon P \to \text{End}(V_k)$ such that $x_{ij} \mapsto t_{ij}$ and $y_{ji} \mapsto t'_{ji}$ for every i and j. Then I is contained in the kernel of φ and $\varphi(1) = 1 \neq 0$, so that $1 \notin I$. This shows that I is a proper ideal of P. It follows that $R := P/I$ is a non-zero k-algebra. If $f \in P$, set $\overline{f} := f + I \in P/I$. Then $\overline{X} := (\overline{x_{ij}})_{i,j} \in M_{n \times m}(R)$ and $\overline{Y} := (\overline{y_{ji}})_{j,i} \in M_{m \times n}(R)$. Since $\overline{XY} = 1_n$, $\overline{YX} = 1_m$, the mappings
$$\overline{X} \colon R_R^m \to R_R^n, \quad \overline{Y} \colon R_R^n \to R_R^m$$
are one the inverse of the other, so that $R_R^n \cong R_R^m$. The k-algebra R is called the *Leavitt algebra* of type (n, m).

2.5 Exact Sequences

From now on, all modules will be right modules over a fixed ring R. We will denote the zero module with one element by 0.

We will now consider *sequences of modules*, where by a sequence of modules
$$\ldots \to M_{i-1} \xrightarrow{f_{i-1}} M_i \xrightarrow{f_i} M_{i+1} \xrightarrow{f_{i+1}} \ldots,$$
we mean a family of modules M_i indexed by integer numbers and a family of module morphisms $f_i \colon M_i \to M_{i+1}$. Sequences can be either finite or infinite on one side or both sides.

A sequence of modules is called a 0-*sequence* (or *a complex of modules*, or *a chain complex of modules*) if $f_i(M_i) \subseteq \ker f_{i+1}$ for every index i. Equivalently, if $f_{i+1} f_i = 0$ for every i.

A sequence is *exact in* M_i if $f_{i-1}(M_{i-1}) = \ker f_i$. A sequence is *exact* if it is exact in M_i for every i.

Examples 2.5.1
Let $f \colon M \to N$ be any module homomorphism.
(1) The sequence $0 \to M \xrightarrow{f} N$ is exact if and only if f is injective.
(2) The sequence $M \xrightarrow{f} N \to 0$ is exact if and only if f is surjective.
(3) A sequence $0 \to M \xrightarrow{f} N \xrightarrow{g} P \to 0$ is exact if and only if f is injective, g is surjective and $f(M) = \ker g$.

Exact sequences of the form

$$0 \to M \to N \to P \to 0$$

are called *short exact sequences* (s. e. s., for short).

Examples 2.5.2
(a) For every submodule N of M, there is a short exact sequence

$$0 \to N \xrightarrow{\iota} M \xrightarrow{\pi} M/N \to 0,$$

where ι denotes the embedding of N into M and π denotes the canonical projection of M onto N.

(b) For every pair of modules M and N, there is a short exact sequence

$$0 \to M \xrightarrow{\varepsilon} M \oplus N \xrightarrow{\pi} N \to 0,$$

where $\varepsilon(m) = (m, 0)$ and $\pi(m, n) = n$ for every $m \in M$ and $n \in N$.

Lemma 2.5.3
Let $f: M \to N$ and $g: N \to M$ be homomorphisms with $gf = \iota_M$. Then $N = f(M) \oplus \ker g$.

Proof Since $f(M)$ and $\ker g$ are submodules of N, we have that $f(M) + \ker g$ is a submodule of N. Conversely, take any $n \in N$. Then $n = fg(n) + (n - fg(n))$, and obviously $fg(n) \in f(M)$. Moreover, $n - fg(n) \in \ker g$, because

$$g(n - fg(n)) = g(n) - gfg(n) = g(n) - \iota_M g(n) = 0$$

This proves that $N = f(M) + \ker g$.

Let us show that $f(M) \cap \ker g = 0$. If $x \in f(M) \cap \ker g$, then $x = f(m)$ for some $m \in M$, and $g(x) = 0$, so that $m = gf(m) = g(x) = 0$. Thus $x = f(m) = 0$. □

2.5 Exact Sequences

Proposition 2.5.4

The following conditions are equivalent for a short exact sequence $0 \to A \xrightarrow{f} B \xrightarrow{g} C \to 0$:

(i) *There exists a homomorphism $f': B \to A$ such that $f'f = \iota_A$ (i.e., f is left invertible).*
(ii) *There exists a homomorphism $g': C \to B$ such that $gg' = \iota_C$ (i.e., g is right invertible).*
(iii) *$f(A) = \ker g$ is a direct summand of B.*

Moreover, if these three equivalent conditions hold, then $B \cong A \oplus C$.

Proof

(i) $\Rightarrow$ (ii) Let $f': B \to A$ be such that $f'f = \iota_A$. By Lemma 2.5.3, we know that $B = f(A) \oplus \ker f'$. Consider the restriction $g|_{\ker f'}: \ker f' \to C$ of the homomorphism g. This restriction is an injective homomorphism, because $\ker(g|_{\ker f'}) = \ker g \cap \ker f' = f(A) \cap \ker f' = 0$. It is also surjective, because $C = g(B) = g(f(A) + \ker f') = g(\ker g + \ker f') = g(\ker f') = g|_{\ker f'}(\ker f')$. Hence $g|_{\ker f'}$ is an isomorphism, so that it has an inverse homomorphism $(g|_{\ker f'})^{-1}: C \to \ker f'$.

Notice that, in particular, $C \cong \ker f'$ and $A \cong f(A)$ imply that

$$B = f(A) \oplus \ker f' \cong A \oplus C,$$

which proves the last part of the statement of the proposition.

Let $\varepsilon: \ker f' \to B$ be the embedding, so that, $g|_{\ker f'} = g\varepsilon$. Set

$$g' := \varepsilon(g|_{\ker f'})^{-1}: C \to B.$$

Then g' is the required right inverse of g, because

$$gg' = g\varepsilon(g|_{\ker f'})^{-1} = (g|_{\ker f'})(g|_{\ker f'})^{-1} = \iota_C.$$

(ii) $\Rightarrow$ (iii) From $gg' = \iota_C$, we have that $B = g'(C) \oplus \ker g$ by Lemma 2.5.3.

(iii) $\Rightarrow$ (i) Assume that $B = f(A) \oplus B'$ for some submodule B' of B. Let $\pi: B = f(A) \oplus B' \to f(A)$ be the canonical projection with kernel B'. Let $f_0: A \to f(A)$ be the homomorphism obtained from f by restricting the image to $f(A)$ (f_0 is some times called the *corestriction* of f to $f(A)$, denoted $f|^{f(A)}$). Then f_0 is an isomorphism and if $\varepsilon: f(A) \to B$ is the embedding, we have that $f = \varepsilon f_0$. Set $f' := f_0^{-1}\pi: B \to A$. Then f' is the required left inverse of f, because

$$f'f = f_0^{-1}\pi\varepsilon f_0 = f_0^{-1}\iota_{f(A)}f_0 = \iota_A.$$

□

A short exact sequence satisfying the three equivalent conditions in the statement of Proposition 2.5.4 is called a *split* exact sequence.

Example 2.5.5
(A short exact sequence $0 \to A \to B \to C \to 0$ which is not split, although $B \cong A \oplus C$.) Let R be the ring $\mathbb{Z}$, so that R-modules are exactly abelian groups. Set $A := \mathbb{Z}$, let C be a countably dimensional vector space over the field $\mathbb{Z}/2\mathbb{Z}$, so that we can write C in the form $C = \bigoplus_{i=0}^{+\infty}(\mathbb{Z}/2\mathbb{Z})v_i$, where $\{v_i \mid i \geq 0\}$ is a basis of the $\mathbb{Z}/2\mathbb{Z}$-vector space C. Set $B := \mathbb{Z} \oplus (\bigoplus_{i=0}^{+\infty}(\mathbb{Z}/2\mathbb{Z})v_i)$.

Consider the exact sequence

$$0 \to A \xrightarrow{f} B \xrightarrow{g} C \to 0, \tag{2.1}$$

where $f \colon \mathbb{Z} \to \mathbb{Z} \oplus (\bigoplus_{i=0}^{+\infty}(\mathbb{Z}/2\mathbb{Z})v_i)$ is defined by $f(z) = (2z, \overline{0}, \overline{0}, \ldots)$ (here $\overline{t} = t + 2\mathbb{Z}$ for every $t \in \mathbb{Z}$). The image of f is $2\mathbb{Z} \oplus 0$. Let

$$g \colon \mathbb{Z} \oplus \left(\bigoplus_{i=0}^{+\infty}(\mathbb{Z}/2\mathbb{Z})v_i\right) \to \bigoplus_{i=0}^{+\infty}(\mathbb{Z}/2\mathbb{Z})v_i$$

be the mapping defined by

$$g \colon (z, \overline{a_0}, \overline{a_1}, \overline{a_2}, \ldots) \mapsto (\overline{z}, \overline{a_0}, \overline{a_1}, \ldots).$$

Then $\ker g = 2\mathbb{Z} \oplus 0$, hence (2.1) is a short exact sequence.

Assume that (2.1) splits, so that there exists a homomorphism

$$f' \colon \mathbb{Z} \oplus \left(\bigoplus_{i=0}^{+\infty}(\mathbb{Z}/2\mathbb{Z})v_i\right) \to \mathbb{Z}$$

with $f'f = \iota_{\mathbb{Z}}$. Then there exist homomorphisms

$$f_1 \colon \mathbb{Z} \to \mathbb{Z} \quad \text{and} \quad f_2 \colon \bigoplus_{i=0}^{+\infty}(\mathbb{Z}/2\mathbb{Z})v_i \to \mathbb{Z}$$

such that $f'(z, v) = f_1(z) + f_2(v)$. For all $t \in \mathbb{Z}$, we have that $t = \iota_{\mathbb{Z}}(t) = f'f(t) = f'(2t, \overline{0}, \overline{0}, \ldots) = f_1(2t) + f_2(\overline{0}, \overline{0}, \ldots) = f_1(2t)$. Setting $t = 1$, we obtain $1 = f_1(2) = 2f_1(1) \in 2\mathbb{Z}$, which implies that 1 is even, a contradiction.

2.6 Projective Modules

Definition 2.6.1

Let R be a ring. A right R-module P_R is *projective* if for every epimorphism $f \colon M_R \to N_R$ and every homomorphism $g \colon P_R \to N_R$, there exists a morphism $h \colon P_R \to M_R$ with $f \circ h = g$.

2.6 Projective Modules

The situation in the previous definition is described by the following commutative diagram, in which the row is exact:

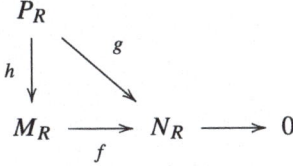

Lemma 2.6.2
(i) *Every free module is projective.*
(ii) *Every direct summand of a projective module is projective.*
(iii) *Every direct sum of projective modules is projective.*

Proof
(i) Let F_R be a free module. Then F_R has a free set of generators X. Let $f: M_R \to N_R$ be an epimorphism and $g: F_R \to N_R$ be an arbitrary homomorphism. For every $x \in X$, let m_x be an element of M_R with $f(m_x) = g(x)$. Set $\overline{f}: X \to M_R$, $x \mapsto m_x$. By the universal property of free R-modules, there exists a unique module homomorphism $h: F_R \to M_R$ that extends $\overline{f}$. Then h is the required homomorphism, because, for every $x \in X$, $f(h(x)) = f(\overline{f}(x)) = f(m_x) = g(x)$. Therefore $f \circ h = g$.
(ii) Let P_R be a projective module and assume $P = A \oplus B$. We will show that A is projective. Let $f: M_R \to N_R$ be an epimorphism and $g: A_R \to N_R$ be a homomorphism. Let $\varepsilon: A \to P$ and $\pi: P \to A$ be the embedding and the canonical projection, so that $\pi \circ \varepsilon = \iota_A$. Since P is projective, there exists $h: P \to M$ with $f \circ h = g \circ \pi$, so that $f \circ h \circ \varepsilon = g \circ \pi \circ \varepsilon = g \circ \iota_A = g$. Thus A is projective.
(iii) is left as an exercise. □

Proposition 2.6.3
The following conditions are equivalent for a right R-module P_R:

(i) *The module P_R is projective.*
(ii) *Every short exact sequence of the form $0 \to M_R \to N_R \to P_R \to 0$ splits.*
(iii) *The module P_R is isomorphic to a direct summand of a free module.*

Proof
(i) ⇒ (ii)   Let P_R be a projective module and let

$$0 \to M_R \to N_R \xrightarrow{g} P_R \to 0 \qquad (2.2)$$

be a short exact sequence. As g is onto and P_R is projective, there exists a morphism $h\colon P_R \to N_R$ such that the diagram

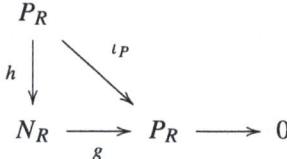

commutes. Hence g is right invertible. Thus (2.2) splits by Proposition 2.5.4(ii).

(ii) ⇒ (iii)   Assume that (ii) holds. Since every R-module is a homomorphic image of a free module, there exist a free R-module F_R and an epimorphism $g\colon F_R \to P_R$. Thus we have an exact sequence

$$0 \to \ker(g) \hookrightarrow F_R \xrightarrow{g} P_R \to 0.$$

This exact sequence splits by (ii), so that $F_R \cong P_R \oplus \ker(g)$.

(iii) ⇒ (i)   Every free module is projective, and every direct summand of a projective module is projective (Lemma 2.6.2, (i) and (ii)). □

Thus projective modules are exactly the modules isomorphic to direct summands of free modules. Since every free module is projective and every module is a homomorphic image of a free module (Proposition 2.2.1(i)), we get that every module is a homomorphic image of a projective module. Similarly, every finitely generated module is a homomorphic image of a finitely generated projective module (Proposition 2.2.1(ii)).

Corollary 2.6.4
A module P_R is a finitely generated projective module if and only if it is isomorphic to a direct summand of R_R^n for some $n \geq 0$.

2.7 Tensor Product of Modules

Proof If P_R is a finitely generated projective module and $\{p_1, \ldots, p_n\}$ is a finite set of generators of P_R, then the mapping

$$\varphi: R_R^n \to P_R, \quad (r_1, \ldots, r_n) \mapsto \sum_{i=1}^{n} p_i r_i$$

is an epimorphism. Hence there is an exact sequence $0 \to \ker(\varphi) \hookrightarrow R_R^n \xrightarrow{\varphi} P_R \to 0$. As P_R is projective, this exact sequence splits by Proposition 2.6.3 ((i) $\Rightarrow$(ii)). Hence $R_R^n \cong P_R \oplus \ker(\varphi)$.

Conversely, if a module P_R is a direct summand of R_R^n, then P_R is projective by Proposition 2.6.3((iii) $\Rightarrow$(i)) and is finitely generated because every homomorphic image of a finitely generated module is finitely generated. □

We will not prove the following interesting result, due to Kaplansky [16].

Theorem 2.6.5
Let R be a ring and M_R an R-module which is a direct sum of a family of countably generated right R-modules. Then any direct summand of M is also a direct sum of a family of countably generated right R-modules.

As a corollary, Kaplansky proves that:

Theorem 2.6.6 ([16])
Every projective module is a direct sum of countably generated projective modules.

2.7 Tensor Product of Modules

Definition 2.7.1

Let R be a ring, M_R a right R-module, $_RN$ a left R-module and G an additive abelian group. A mapping $\beta: M_R \times {}_RN \to G$ is said to be a *balanced mapping* (or an *R-balanced mapping*) if, for every $m, m' \in M_R, n, n' \in {}_RN$ and $r \in R$,

(i) $\beta(m + m', n) = \beta(m, n) + \beta(m', n)$.
(ii) $\beta(m, n + n') = \beta(m, n) + \beta(m, n')$.
(iii) $\beta(mr, n) = \beta(m, rn)$.

If $\beta\colon M_R \times {}_R N \to G$ is a balanced mapping, then, for any fixed $m \in M_R$, the mapping $\beta(m, -)\colon N \to G$, defined by $n \mapsto \beta(m, n)$ for every $n \in N$, is a group homomorphism. It follows that we have that $\beta(m, 0) = 0$, $\beta(m, -n) = -\beta(m, n)$ and $\beta(m, zn) = z\beta(m, n)$ for every $m \in M$, $n \in N$ and $z \in \mathbb{Z}$. Similarly, $\beta(0, n) = 0$, $\beta(-m, n) = -\beta(m, n)$ and $\beta(zm, n) = z\beta(m, n)$ for every $m \in M$, $n \in N$ and $z \in \mathbb{Z}$.

For instance, for every ring R, the multiplication $R_R \times {}_R R \to R$, $(r, s) \mapsto rs$ is a balanced mapping.

Definition 2.7.2

A *tensor product* of a right R-module M_R and a left R-module ${}_R N$ is a pair (T, τ), where T is an abelian group, $\tau\colon M_R \times {}_R N \to T$ is a balanced mapping, and the following universal property holds: for any abelian group G and any balanced mapping $\beta\colon M_R \times {}_R N \to G$ there exists a unique group homomorphism $\widetilde{\beta}\colon T \to G$ such that $\beta = \widetilde{\beta} \circ \tau$, that is, such that the diagram

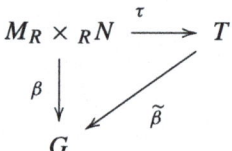

commutes.

Thus a pair (T, τ), where T is an abelian group and $\tau\colon M_R \times {}_R N \to T$ is a balanced mapping, is a tensor product of M_R and ${}_R N$ if and only if for every abelian group G the mapping

$$\mathrm{Hom}_{\mathbf{Ab}}(T, G) \to \mathrm{Bal}(M_R \times {}_R N, G)$$
$$f \mapsto f \circ \tau$$

is a bijection. Here $\mathrm{Bal}(M_R \times {}_R N, G)$ denotes the set of all balanced mappings $M_R \times {}_R N \to G$.

Proposition 2.7.3
Given any ring R and any two modules M_R and ${}_R N$, a tensor product (T, τ) of M_R and ${}_R N$ always exists.

Proof Let

$$\mathbb{Z}^{(M \times N)} := \left\{ \sum_{i=1}^n z_i(m_i, n_i) \mid z_i \in \mathbb{Z},\ m_i \in M,\ n_i \in N,\ n \geq 0 \right\}$$

2.7 Tensor Product of Modules

be the free $\mathbb{Z}$-module with $M \times N$ as a free set of generators. Let S be the subgroup of $\mathbb{Z}^{(M \times N)}$ generated by all the elements of one of the following forms:

1. $(m + m', n) - (m, n) - (m', n)$ for any $m, m' \in M$, $n \in N$.
2. $(m, n + n') - (m, n) - (m, n')$ for any $m \in M$, $n, n' \in N$.
3. $(mr, n) - (m, rn)$ for any $m \in M$, $n \in N$, $r \in R$.

Set
$$T := \mathbb{Z}^{(M \times N)}/S$$
and
$$\tau \colon M \times N \to T, \ (m, n) \mapsto (m, n) + S$$

By construction, τ is balanced. We have to show that the pair (T, τ) has the required universal property. Let G be an abelian group, and let $\beta \colon M \times N \to G$ be a balanced mapping. We must show that there exists a unique group homomorphism $\widetilde{\beta} \colon T \to G$ such that $\beta = \widetilde{\beta} \circ \tau$.

Existence of $\widetilde{\beta}$. The mapping $\beta \colon M \times N \to G$ extends uniquely to a group homomorphism $\beta' \colon \mathbb{Z}^{(M \times N)} \to G$ (universal property of free modules). Since β is balanced, we have that $\beta'(S) = 0$, because β' sends the generators of S to zero. Therefore β' induces a group morphism $\widetilde{\beta} \colon T \to G$. Moreover, $\widetilde{\beta}(\tau(m, n)) = \widetilde{\beta}((m, n) + S) = \beta'(m, n) = \beta(m, n)$ for any $(m, n) \in M \times N$. So that $\widetilde{\beta} \circ \tau = \beta$.

Uniqueness of $\widetilde{\beta}$. Take another group morphism $\alpha \colon T \to G$ such that $\alpha \circ \tau = \beta$. We must show that $\alpha = \widetilde{\beta}$. Now $\alpha \circ \tau = \beta = \widetilde{\beta} \circ \tau$, so that α and $\widetilde{\beta}$ coincide on all elements of the form $\tau(m, n) = (m, n) + S$. But $M \times N$ is a set of generators for $\mathbb{Z}^{(M \times N)}$, so that $\{\tau(m, n) \mid m \in M, n \in N\}$ is a set of generators for T. Then $\alpha = \widetilde{\beta}$ on T. □

Proposition 2.7.4 (Uniqueness of the Tensor Product Up to Isomorphism)
Let R be a ring, M_R a right R-module, $_RN$ a left R-module, and $(T, \tau), (T', \tau')$ two tensor products of M_R and $_RN$. Then there is a unique group isomorphism $\varphi \colon T \to T'$ for which the diagram

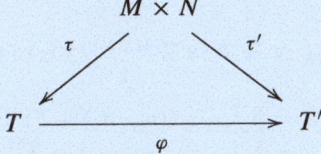

commutes.

Proof Apply the universal property of the tensor product (T, τ) to the balanced mapping $\tau'\colon M \times N \to T'$, so that $\tau' = \varphi \circ \tau$ for a unique group homomorphism $\varphi\colon T \to T'$. It remains to prove only that φ is an isomorphism. Apply the universal property of the tensor product (T', τ') to the balanced mapping $\tau\colon M \times N \to T$, so that $\tau = \varphi' \circ \tau'$ for a unique group homomorphism $\varphi'\colon T' \to T$. We have

$$\tau = \varphi' \circ \tau' = \varphi' \circ \varphi \circ \tau$$

$$\tau' = \varphi \circ \tau = \varphi \circ \varphi' \circ \tau'.$$

From the universal property of (T, τ) applied to the balanced mapping $\tau\colon M \times N \to T$, we know that there exists a unique group homomorphism $f\colon T \to T$ such that $\tau = f \circ \tau$. Both the identity ι_T and $\varphi' \circ \varphi$ have this property, because $\iota_T \circ \tau = \tau = \varphi \circ \varphi' \circ \tau'$. By the uniqueness that is part of the universal property, $\varphi' \circ \varphi = \iota_T$. Similarly, apply the universal property of (T', τ') to the balanced mapping $\tau'\colon M \times N \to T'$, getting that there exists a unique group homomorphism $g\colon T' \to T'$ such that $\tau' = g \circ \tau'$. Both the identity $\iota_{T'}$ and $\varphi' \circ \varphi$ have this property. By the uniqueness, $\varphi \circ \varphi' = \iota_{T'}$. Hence φ is an isomorphism. □

Given a ring R and two modules M_R and $_R N$, if (T, τ) is a tensor product of M and N, then T will be denoted by $M \otimes_R N$ and $\tau(m, n)$ by $m \otimes n$ for every $(m, n) \in M \times N$.

Since τ is balanced, in this notation we have, for every $m \in M$, $n \in N$ and $r \in R$, that $(m + m') \otimes n = m \otimes n + m' \otimes n$, $m \otimes (n + n') = m \otimes n + m \otimes n'$, $(mr) \otimes n = m \otimes (rn)$, $m \otimes 0 = 0$, $m \otimes (-n) = -(m \otimes n)$, etc.

Proposition 2.7.5
Any element of $M \otimes_R N$ can be written in the form

$$\sum_{i=1}^{t} m_i \otimes n_i$$

for suitable $m_i \in M$, $n_i \in N$, $t \geq 1$.

Proof Any element of $M \otimes_R N = T = \mathbb{Z}^{(M \times N)}/S$ is of the form

$$\left(\sum_{i=1}^{t} z_i(m_i, n_i)\right) + S = \sum_{i=1}^{t} z_i((m_i, n_i) + S) = \sum_{i=1}^{t} z_i \tau(m_i, n_i)$$
$$= \sum_{i=1}^{t} \tau(z_i m_i, n_i) = \sum_{i=1}^{t} (z_i m_i) \otimes n_i.$$

It is now sufficient to call m_i the elements $z_i m_i$. □

Tensor Product of Morphisms

Let $\varphi\colon M_R \to M'_R$ be a right R-module morphism and $\psi\colon {}_RN \to {}_R N'$ a left R-module morphism. Define

$$\omega\colon M_R \times {}_RN \to M' \otimes_R N'$$

by $\omega(x,y) = \varphi(x) \otimes \psi(y)$ (We are writing both the mappings φ and ψ on the left). It is easily verified that ω is a balanced mapping, so that there is a unique group morphism

$$\varphi \otimes \psi \colon M \otimes_R N \to M' \otimes_R N'$$

such that $(\varphi \otimes \psi)(x \otimes y) = \varphi(x) \otimes \psi(y)$ for every $x \in M_R$, $y \in {}_RN$. In other words, $\varphi \otimes \psi$ is the unique group morphism making the diagram

$$\begin{array}{ccc} M \times N & \xrightarrow{\varphi \times \psi} & M' \times N' \\ {\scriptstyle \tau}\downarrow & & \downarrow{\scriptstyle \tau'} \\ M \otimes_R N & \xrightarrow{\varphi \otimes \psi} & M' \otimes_R N' \end{array}$$

commute. The group morphism $\varphi \otimes \psi$ is called the *tensor product* of φ and ψ.

Let

$$\varphi\colon M_R \to M'_R, \quad \varphi'\colon M'_R \to M''_R$$

be right R-module homomorphisms and let

$$\psi\colon {}_RN \to {}_RN', \quad \psi'\colon {}_RN' \to {}_RN''$$

be left R-module homomorphisms. It is easily checked that $(\varphi' \circ \varphi) \otimes (\psi' \circ \psi)(x \otimes y) = (\varphi' \otimes \psi') \circ (\varphi \otimes \psi)(x \otimes y)$, that is, the group morphisms $(\varphi' \circ \varphi) \otimes (\psi' \circ \psi)$ and $(\varphi' \otimes \psi') \circ (\varphi \otimes \psi)$ agree on the elements of the form $x \otimes y$. Since these elements generate the abelian group $M \otimes_R N$ (Proposition 2.7.5), it follows that $(\varphi' \circ \varphi) \otimes (\psi' \circ \psi) = (\varphi' \otimes \psi') \circ (\varphi \otimes \psi)$. Similarly, one sees that $\iota_M \otimes \iota_N = \iota_{M \otimes N}$.

In particular, if φ and ψ are isomorphisms, then so is $\varphi \otimes \psi$ and its inverse is $\varphi^{-1} \otimes \psi^{-1}$, because

$$(\varphi \otimes \psi) \circ (\varphi^{-1} \otimes \psi^{-1}) = (\varphi \circ \varphi^{-1}) \otimes (\psi \circ \psi^{-1}) = \iota_{M'} \otimes \iota_{N'} = \iota_{M' \otimes N'}.$$

Now fix a right module M_R over a ring R. We can define a functor of the category R-Mod of all left R-modules into the category **Ab** of abelian groups as follows. To every left R-module $_RN$, we associate a tensor product $M \otimes_R N$. To every left R-module homomorphism $\varphi: {}_RN \to {}_RN'$, we associate the group homomorphism $\iota_M \otimes_R \varphi: M \otimes_R N \to M \otimes_R N'$. We will denote this functor as $M \otimes_R -$.

Similarly on the other side. Fix a left module $_RN$ over a ring R. We can define a functor of the category Mod-R into the category **Ab** in the following way. To every right R-module M_R, we associate a tensor product $M \otimes_R N$. To every right R-module homomorphism $\psi: M_R \to M'_R$, we associate the group homomorphism $\psi \otimes_R \iota_N: M \otimes_R N \to M' \otimes_R N$. We will denote this functor by $- \otimes_R N$.

Exercise 2.7.6
These functors are defined only up to a natural isomorphism, in the following sense. Fix a right module M_R. We can define a functor $F: R\text{-Mod} \to \mathbf{Ab}$ associating to every left R-module $_RN$ a tensor product T_N of M_R and $_RN$. We can also define another functor $F': R\text{-Mod} \to \mathbf{Ab}$ associating to every left R-module $_RN$ another tensor product T'_N of M_R and $_RN$. Show that there is a natural isomorphism $\eta: F \to F'$.

Exercise 2.7.7
Compute $\mathbb{Q} \otimes_\mathbb{Z} (\mathbb{Z}/n\mathbb{Z})$, where n is a positive integer.

[Hint: Multiplication by n is an isomorphism $\mathbb{Q} \to \mathbb{Q}$. Tensor it with the identity $\mathbb{Z}/n\mathbb{Z} \to \mathbb{Z}/n\mathbb{Z}$. One obtains an homomorphism $\mathbb{Q} \otimes_\mathbb{Z} \mathbb{Z}/n\mathbb{Z} \to \mathbb{Q} \otimes_\mathbb{Z} \mathbb{Z}/n\mathbb{Z}$, which coincides with multiplication by n. Multiplication by n is the zero homomorphism $\mathbb{Z}/n\mathbb{Z} \to \mathbb{Z}/n\mathbb{Z}$. Tensor it with the identity $\mathbb{Q} \to \mathbb{Q}$. The homomorphism $\mathbb{Q} \otimes_\mathbb{Z} \mathbb{Z}/n\mathbb{Z} \to \mathbb{Q} \otimes_\mathbb{Z} \mathbb{Z}/n\mathbb{Z}$ obtained also coincides with multiplication by n. Other possible approach: any generator $q \otimes \bar{z}$ of $\mathbb{Q} \otimes_\mathbb{Z} (\mathbb{Z}/n\mathbb{Z})$ can be written in the form $\frac{nq}{n} \otimes \bar{z} = \frac{q}{n} \otimes n\bar{z} = \frac{q}{n} \otimes 0 = \dots$]

Exercise 2.7.8
Compute $(\mathbb{Z}/n\mathbb{Z}) \otimes_\mathbb{Z} (\mathbb{Z}/m\mathbb{Z})$, where n and m are coprime positive integers.

[Hint: There exist $a, b \in \mathbb{Z}$ such that $1 = an + bm$. For every $x \in \mathbb{Z}/n\mathbb{Z}$, $y \in \mathbb{Z}/m\mathbb{Z}$, one has $x \otimes y = (an + bm)(x \otimes y) = an(x \otimes y) + bm(x \otimes y) = \dots = 0$.]

Notice that the fact that $M \otimes_R N = 0$ is equivalent to saying that there are no non-zero balanced mappings $\beta: M_R \times {}_RN \to G$ into any abelian group G.

Proposition 2.7.9

Let R, S, T be rings. Assume that $_SM_R$ and $_RN_T$ are S-R- and R-T-bimodules, respectively. Then $_S(M \otimes N)_T$ is a S-T-bimodule in which the multiplications are defined by

$$s\left(\sum_{i=1}^{p} m_i \otimes n_i\right) = \sum_{i=1}^{p}(sm_i) \otimes n_i \quad \text{and} \quad \left(\sum_{i=1}^{p} m_i \otimes n_i\right)t = \sum_{i=1}^{p} m_i \otimes (n_i t)$$

for every $s \in S$, $\sum_{i=1}^{p} m_i \otimes n_i \in M \otimes_R N$ and $t \in T$.

2.7 Tensor Product of Modules

Proof First of all we must check that the left S-module structure and the right T-module structure on $M \otimes_R N$ are well defined. Fix an element $s \in S$ and consider the homomorphism induced by left multiplication by s:

$$\lambda_s \colon M_R \to M_R, \ x \mapsto sx.$$

This λ_s is a right R-module morphism. Thus $\lambda_s \otimes \iota_N \colon M \otimes_R N \to M \otimes_R N$ is a group morphism, and if $s, s' \in S$, $\lambda_{ss'} = \lambda_s \circ \lambda_{s'}$, so that $(\lambda_s \otimes \iota_N) \circ (\lambda_{s'} \otimes \iota_N) = \lambda_{ss'} \otimes \iota_N$. It follows that

$$\Lambda \colon S \to \mathrm{End}(M \otimes_R N)$$
$$s \mapsto \lambda_s \otimes \iota_N$$

is a ring morphism. Hence we get a left S-module structure on $M \otimes_R N$. Moreover, we have:

$$s(\textstyle\sum_{i=1}^p m_i \otimes n_i) = \Lambda(s)(\sum_{i=1}^p m_i \otimes n_i) = (\lambda_s \otimes \iota_N)(\sum_{i=1}^p m_i \otimes n_i)$$
$$= \textstyle\sum_{i=1}^p \lambda_s(m_i) \otimes \iota_N(n_i) = \sum_{i=1}^p (sm_i) \otimes n_i.$$

Similarly, $M \otimes_R N$ is a right T-module in which $\left(\sum_{i=1}^p m_i \otimes n_i\right) t = \sum_{i=1}^p m_i \otimes (n_i t)$. The two multiplications are compatible, because

$$s\left(\left(\textstyle\sum_{i=1}^p m_i \otimes n_i\right) t\right) = s\left(\sum_{i=1}^p m_i \otimes (n_i t)\right) = \sum_{i=1}^p (sm_i) \otimes (n_i t)$$
$$= \left(\textstyle\sum_{i=1}^p (sm_i) \otimes n_i\right) t = \left(s\left(\sum_{i=1}^p m_i \otimes n_i\right)\right) t.$$

Thus ${}_S(M \otimes_R N)_T$ is a S-T-bimodule. □

Corollary 2.7.10

If M and N are modules over a commutative ring R, then $M \otimes_R N$ is an R-module and, for every $r \in R$, $m_i \in M$, $n_i \in N$,

$$r\left(\sum_{i=1}^p m_i \otimes n_i\right) = \sum_{i=1}^p (rm_i) \otimes n_i = \sum_{i=1}^p m_i \otimes (rn_i).$$

Proof Every R-module is an R-R-bimodule (Exercise 1.5.2), hence we can apply the previous Proposition. □

Exercise 2.7.11

Let R be a commutative ring and M, N be modules over R. Show that:
(1) The mapping $\tau \colon M \times N \to M \otimes_R N$ is bilinear.
(2) Show that the pair $(M \otimes_R N, \tau)$ also has the universal property with respect to bilinear mappings: for every R-module P and every bilinear mapping $\beta \colon M \times N \to P$ there exists a unique R-module homomorphism $\widetilde{\beta} \colon M \otimes_R N \to P$ such that $\beta = \widetilde{\beta} \circ \tau$.

Thus the situation over a commutative ring R is the following. Let M and N be modules over a commutative ring R. Then:
(a) For every abelian group G, there is a bijection

$$\operatorname{Hom}_{\mathbf{Ab}}(M \otimes_R N, G) \to R\text{-Bal}(M \times N, G)$$
$$f \in \operatorname{Hom}_{\mathbf{Ab}}(M \otimes_R N, G) \mapsto f \circ \tau.$$

Here Bal denotes balanced mappings.
(b) For every R-module P, there is a bijection

$$\operatorname{Hom}_{R\text{-Mod}}(M \otimes_R N, P) \to R\text{-Bil}(M \times N, P)$$
$$f \in \operatorname{Hom}_{R\text{-Mod}}(M \otimes_R N, P) \mapsto f \circ \tau.$$

Here Bil denotes R-bilinear mappings.
(c) Thus, for every R-module P, we have that

$$\begin{array}{ccc} \operatorname{Hom}_{\mathbf{Ab}}(M \otimes_R N, P) & \to & R\text{-Bal}(M \times N, P) \\ \mathsf{I} \cup & & \mathsf{I} \cup \\ \operatorname{Hom}_{R\text{-Mod}}(M \otimes_R N, P) & \to & R\text{-Bil}(M \times N, P) \end{array}$$

In other words, balanced mappings are all the mappings $M \times N \to G$ of the form $f \circ \tau$ for some group morphism $f \colon M \otimes_R N \to G$ for some abelian group G, and R-bilinear mappings are all the mappings $M \times N \to P$ of the form $g \circ \tau$ for some R-module morphism $g \colon M \otimes_R N \to P$ for some R-module P. For instance, show that for $R = M = N = \mathbb{R}$, there is a surjective $\mathbb{R}$-balanced mapping $\mathbb{R} \times \mathbb{R} \to \mathbb{Q}$ and that all $\mathbb{R}$-bilinear mappings $\beta \colon \mathbb{R} \times \mathbb{R} \to \mathbb{R}$ are of the form $\beta(x, y) = rxy$ for some $r \in \mathbb{R}$.

Exercise 2.7.12

In this exercise, we will prove that "the functor Hom is left exact and the functor $\otimes$ is right exact". We will be more precise on the meaning of "left exact, right exact" in the sequel.
(i) Prove that, for every fixed module M_R, if the sequence of right R-modules $0 \to N'_R \to N_R \to N''_R$ is exact, then so is the sequence of abelian groups

$$0 \to \operatorname{Hom}(M_R, N'_R) \to \operatorname{Hom}(M_R, N_R) \to \operatorname{Hom}(M_R, N''_R).$$

(ii) Prove that, for every fixed module N_R, if the sequence of right R-modules $M'_R \to M_R \to M''_R \to 0$ is exact, then so is the sequence of abelian groups

$$0 \to \operatorname{Hom}(M''_R, N_R) \to \operatorname{Hom}(M_R, N_R) \to \operatorname{Hom}(M'_R, N_R).$$

2.7 Tensor Product of Modules

(iii) Prove that, for every fixed module $_RN$, if the sequence of right R-modules $M'_R \to M_R \to M''_R \to 0$ is exact, then so is the sequence of abelian groups

$$M' \otimes_R N \to M \otimes_R N \to M'' \otimes_R N \to 0.$$

(iv) Prove that, for every fixed module M_R, if the sequence of left R-modules $_RN' \to {_RN} \to {_RN''} \to 0$ is exact, then so is the sequence of abelian groups

$$M \otimes_R N' \to M \otimes_R N \to M \otimes_R N'' \to 0.$$

[Hint for (iii): Let $M'_R \xrightarrow{\alpha} M_R \xrightarrow{\beta} M''_R \to 0$ be an exact sequence. In order to prove that $\ker(\beta \otimes \iota_N)$ is contained in the image of $\alpha \otimes \iota_N$, let $I := (\alpha \otimes \iota_N)(M' \otimes_R N)$ denote the image of $\alpha \otimes \iota_N$ and let $\gamma: M'' \to M$ be a right inverse of the onto mapping $\beta: M \to M''$ (such a mapping γ always exists, though it is not necessarily a module morphism). Define a mapping $b: M'' \times N \to (M \otimes_R N)/I$ by $b(m'', n) = (\gamma(m'') \otimes n) + I$. It is easily seen that b is balanced. Therefore b induces a group morphism $\tilde{b}: M'' \otimes_R N \to (M \otimes_R N)/I$ with the property that $\tilde{b}(m'' \otimes n) = (\gamma(m'') \otimes n) + I$. Thus $\ker(\beta \otimes \iota_N) \subseteq I$, because if $x \in \ker(\beta \otimes \iota_N)$, then ...].

Exercise 2.7.13
In general, the functors $\mathrm{Hom}(M_R, -): \mathrm{Mod}\text{-}R \to \mathbf{Ab}$, $\mathrm{Hom}(-, N_R): \mathrm{Mod}\text{-}R \to \mathbf{Ab}$, $M \otimes_R -: R\text{-Mod}$ and $- \otimes_R N': \mathrm{Mod}\text{-}R \to \mathbf{Ab}$ are not exact, that is, they do not necessarily send a short exact sequence $0 \to A \to B \to C \to 0$ to a short exact sequence.
(i) Show that the functor $\mathrm{Hom}(M_R, -): \mathrm{Mod}\text{-}R \to \mathbf{Ab}$ is exact, that is, for every short exact sequence $0 \to A_R \to B_R \to C_R \to 0$, the sequence $0 \to \mathrm{Hom}(M_R, A_R) \to \mathrm{Hom}(M_R, B_R) \to \mathrm{Hom}(M_R, C_R) \to 0$ is exact, if and only if the module M_R is projective.
(ii) Let R and S be rings and $F: \mathrm{Mod}\text{-}R \to \mathrm{Mod}\text{-}S$ be a covariant functor. Show that the following conditions are equivalent:

(a) The functor F is exact, that is, for every short exact sequence $0 \to A_R \to B_R \to C_R \to 0$ the sequence $0 \to F(A_R) \to F(B_R) \to F(C_R) \to 0$ is exact.
(b) For every exact sequence $A_R \to B_R \to C_R$ the sequence $F(A_R) \to F(B_R) \to F(C_R)$ is exact.
(c) For every exact sequence $\ldots \to M_{i-1} \xrightarrow{f_{i-1}} M_i \xrightarrow{f_i} M_{i+1} \xrightarrow{f_{i+1}} \ldots$ the sequence $\ldots \to F(M_{i-1}) \xrightarrow{F(f_{i-1})} F(M_i) \xrightarrow{F(f_i)} F(M_{i+1}) \xrightarrow{F(f_{i+1})} \ldots$ is exact.

(This explains why short exact sequences are important.)
[Hint for (c) $\Rightarrow$ (a): First show that F assigns to the zero R-module the zero S-module. To this aim, apply the functor F to the exact sequence $0 \xrightarrow{1} 0 \xrightarrow{1} 0$.]

Exercise 2.7.14
Prove that, for every module M_R, the abelian groups $M \otimes_R R$ and M are canonically isomorphic. More precisely, prove that there is a natural isomorphism of the functor $- \otimes_R R: \mathrm{Mod}\text{-}R \to \mathbf{Ab}$ onto the forgetful functor $F: \mathrm{Mod}\text{-}R \to \mathbf{Ab}$.

Exercise 2.7.15
Prove that, for every module M_R and every left ideal I of R, the abelian groups $M \otimes_R R/I$ and M/MI are canonically isomorphic.

More precisely, prove that there is a natural isomorphism of the functor $- \otimes_R R/I \colon \text{Mod-}R \to \textbf{Ab}$ onto the functor $(-/-I) \colon \text{Mod-}R \to \textbf{Ab}$, where $(-/-I)$ is the functor that maps any right R-module M_R to the abelian group M/MI.

Exercise 2.7.16
Let $\{ M_i \mid i \in I \}$ be a family of right R-modules and $\oplus_{i \in I} M_i$ be its direct sum. Let $_R N$ be a left module. Show that $(\oplus_{i \in I} M_i) \otimes_R N \cong \oplus_{i \in I} (M_i \otimes_R N)$ in a canonical way. More precisely, for every $j \in I$, let $\varepsilon_j \colon M_j \to \oplus_{i \in I} M_i$ be the canonical monomorphism. It induces a group homomorphism $\varepsilon_j \otimes N \colon M_j \otimes_R N \to (\oplus_{i \in I} M_i) \otimes_R N$. The family of morphisms $\{ \varepsilon_i \otimes N \mid i \in I \}$ defines a group morphism $(\varepsilon_i \otimes N)_{i \in I} \colon \oplus_{i \in I} (M_i \otimes_R N) \to (\oplus_{i \in I} M_i) \otimes_R N$. Show that this group morphism is an isomorphism.

2.8 Projective $\mathbb{Z}$-Modules: Hereditary Rings

Recall that a *principal ideal domain* (PID) is a commutative integral domain in which every ideal is principal, that is, can be generated by one element. A *right hereditary ring* is a ring in which every right ideal is projective. Thus every PID is a (right) hereditary ring.

> **Theorem 2.8.1**
> *Let R be a right hereditary ring. Then every submodule of a free right R-module is isomorphic to a direct sum of right ideals of R.*

In the next proof we will make use of the "Well-ordering Theorem", according to which every set X can be given a *well-order*, that is, it is always possible to define on X a partial order $\leq$ with the property that every non-empty subset Y of X has a least element (i.e., for every such Y there exists an element $y_0 \in Y$ such that $y_0 \leq y$ for all $y \in Y$.)

Proof Let F be a free right R-module with a free set X of generators. By the well-ordering theorem, we can suppose that the elements of X are indexed in an initial segment of ordinals, that is, $X = \{ x_\alpha \mid \alpha < \tau \}$ for some ordinal τ. For $\sigma \leq \tau$, set $F_\sigma := \oplus_{\alpha < \sigma} x_\alpha R$, so that $F_\tau = F$. Let N be a submodule of F. Set $N_\sigma := N \cap F_\sigma$ for every $\sigma \leq \tau$, so that the N_σ's form an ascending chain of submodules of N whose last element is N. Then $N_{\sigma+1}/N_\sigma$ is projective for every $\sigma < \tau$, because

$$N_{\sigma+1}/N_\sigma = N_{\sigma+1}/(N_{\sigma+1} \cap F_\sigma) \cong (F_\sigma + N_{\sigma+1})/F_\sigma \leq F_{\sigma+1}/F_\sigma \cong R_R.$$

2.8 Projective ℤ-Modules: Hereditary Rings

It follows that the exact sequence $0 \to N_\sigma \to N_{\sigma+1} \to N_{\sigma+1}/N_\sigma \to 0$ splits, i.e., $N_{\sigma+1} = N_\sigma \oplus C_\sigma$ for a suitable submodule C_σ of $N_{\sigma+1}$. Notice that $C_\sigma \cong N_{\sigma+1}/N_\sigma$ is isomorphic to a right ideal of R. We will show that $N = \oplus_{\sigma < \tau} C_\sigma$.

Set $S := \sum_{\sigma < \tau} C_\sigma$. Clearly, $N \supseteq S$. If $N \not\subseteq S$, there is a smallest index α with $N_\alpha \not\subseteq S$. If α is a limit ordinal, then $N_\alpha = \cup_{\beta < \alpha} N_\beta$, and $N_\beta \subseteq S$ for every $\beta < \alpha$, contradiction. Thus α is not a limit ordinal, that is, $\alpha = \alpha' + 1$ for some ordinal α', and $N_{\alpha'} \subseteq S$. But then $N_\alpha = N_{\alpha'} \oplus C_{\alpha'} \subseteq S$, contradiction. This contradiction shows that $N \subseteq S$.

In order to prove that the sum S is direct, assume $c_{\sigma_1} + c_{\sigma_2} + \cdots + c_{\sigma_n} = 0$ with c_{σ_i} non-zero element of C_{σ_i}, and $\sigma_1 < \sigma_2 < \cdots < \sigma_n$. Then $c_{\sigma_n} = -c_{\sigma_1} - c_{\sigma_2} - \cdots - c_{\sigma_{n-1}} \in C_{\sigma_n} \cap (C_{\sigma_1} + C_{\sigma_2} + \cdots + C_{\sigma_{n-1}}) \subseteq C_{\sigma_n} \cap N_{\sigma_n} = 0$, a contradiction. This shows that the sum is direct, as desired. □

Corollary 2.8.2
Let R be a PID. Then every submodule of a free R-module is free. In particular, projective R-modules are free.

Proof A PID is hereditary, so that by the previous theorem, every submodule of a free R-module is isomorphic to a direct sum of right ideals of R. Now the ideals of R are R-modules that are either isomorphic to R or 0. In any case, a direct sum of ideals of R is a free module. In particular, projective R-modules are isomorphic to direct summands of free modules, so that they are isomorphic to submodules of free modules, and therefore they are free. □

This corollary applies in particular to ℤ-modules, that is, to abelian groups, so that the class of projective ℤ-modules coincides with the class of free ℤ-modules.

We will give in Exercise 3.1.6 an example of a ring that is right hereditary but not left hereditary. Hereditary commutative integral domains are called *Dedekind domains*. More precisely, it would be possible to prove that:

Proposition 2.8.3
The following conditions are equivalent for a commutative integral domain R:

(i) *Every non-zero proper ideal of R is a product of prime ideals.*
(ii) *Every non-zero proper ideal of R is a product of prime ideals in a unique way up to the order of the factors.*
(iii) *Every ideal of R is a projective R-module.*
(iv) *Every submodule of a projective R-module is a projective R-module.*
(v) *Every submodule of a free R-module is a projective R-module.*

Dedekind domains are the commutative integral domains that satisfy the equivalent conditions of this proposition.

2.9 Idempotents, Nilpotents, Further Results on Projective Modules

If R is a ring and I, J are two right ideals of R, the product of I and J is by definition

$$IJ := \{ \sum_{k=1}^{n} i_k j_k \mid n \geq 1, \ i_k \in I, \ j_k \in J \ \text{for every } k = 1, 2, \ldots, n \}. \tag{2.3}$$

It is a right ideal of R.

Let R be a ring, $r \in R$ and I a right ideal of R. We say that

r is *idempotent* if $r^2 = r$;
r is *nilpotent* if $r^n = 0$ for some $n \geq 1$;
I is *idempotent* if $I^2 = I$;
I is *nil* if every element of I is nilpotent;
I is *nilpotent* if $I^n = 0$ for some $n \geq 1$.

▶ **Remarks 2.9.1**

(1) If $r \in R$ is idempotent, then $r^n = r$ for every $n \geq 1$.
(2) A right ideal I is nilpotent if and only if there exists an integer $n \geq 1$ such that $a_1 \ldots a_n = 0$ for every $a_1, \ldots, a_n \in I$.
(3) Every nilpotent right ideal is nil.

Example 2.9.2
Consider the commutative ring

$$R = k[X_1, X_2, X_3, \ldots]/(X_1, X_2^2, X_3^3, \ldots),$$

where k is a field. We will show that

$$I = (X_1, X_2, X_3, \ldots)/(X_1, X_2^2, X_3^3, \ldots)$$

is a nil ideal of R, but not a nilpotent ideal.

In order to prove that I is nil, consider an arbitrary element

$$f + (X_1, X_2^2, X_3^3, \ldots)$$

of I. Here f is an element of the ideal $(X_1, X_2, X_3, \ldots)$ of $k[X_1, X_2, X_3, \ldots]$, so that $f = \sum_{i=1}^{n} X_i f_i$ for some $n \geq 1$ and suitable polynomials

$$f_i \in k[X_1, X_2, X_3, \ldots].$$

Then $f^{1+2+\cdots+n} = \left(\sum_{i=1}^{n} X_i f_i \right)^{1+2+\cdots+n} \in (X_1, X_2^2, X_3^3, \ldots)$.

2.9 Idempotents, Nilpotents, Further Results on Projective Modules

In order to show that I is not nilpotent, assume the contrary, so that there exists N with $I^N = 0$. Then $(X_1, X_2, X_3, \ldots)^N \subseteq (X_1, X_2^2, X_3^3, \ldots)$, so that in particular $X_{N+1}^N \in (X_1, X_2^2, X_3^3, \ldots)$. It follows that there exist $n \geq 1$ and $f_1, \ldots, f_n \in k[X_1, X_2, X_3, \ldots]$ with $X_{N+1}^N = \sum_{i=1}^n X_i^i f_i$. Here we can assume $n \geq N+1$ without loss of generality. Now apply to the equality $X_{N+1}^N = \sum_{i=1}^n X_i^i f_i$ the homomorphism $k[X_1, X_2, X_3, \ldots] \to k[X]$ that maps X_i to $\delta_{i,N+1} X$. This means that we orderly replace the tuple

$$(X_1, X_2, X_3, \ldots) \quad \text{with the tuple} \quad (0, 0, \ldots, 0, \underset{\underset{N+1}{\uparrow}}{X}, 0, \ldots).$$

We get the equality $X^N = X^{N+1} f$ for some $f \in k[X]$, which is not possible because X^N has degree N, while $X^{N+1} f$ has either degree $> N$ or is 0.

Proposition 2.9.3
Let R be a ring and I, J right ideals of R. Then $R_R = I \oplus J$ if and only if there exists an idempotent element $e \in R$ with $I = eR$ and $J = (1-e)R$.

Proof
($\Rightarrow$) Assume that $R_R = I \oplus J$. Then $1 = e + f$ for some unique $e \in I$, $f \in J$. If i is any element of I, then $i = 1 \cdot i = (e+f)i = ei + fi$, whence $i - ei = fi \in I \cap J$. Thus $i - ei = 0$, so that in particular $e = e^2$ is idempotent, and $i \in eR$. This proves that $I \subseteq eR$. Conversely, $eR \subseteq I$ because $e \in I$. Similarly, $J = fR = (1-e)R$.

($\Leftarrow$) Let $e \in R$ be an idempotent element. We want to show that $R_R = eR \oplus (1-e)R$. Clearly, $R_R \supseteq eR + (1-e)R$. Conversely, fix $r \in R$. Then $r = 1 \cdot r = (e + (1-e))r = er + (1-e)r \in eR + (1-e)R$. Thus $R_R = eR + (1-e)R$. It remains to show that the sum is direct. Fix $x \in eR \cap (1-e)R$. Then $x = er = (1-e)s$ for suitable $r, s \in R$, and $x = er = eer = e(1-e)s = 0$. □

Hence, if $e \in R$ is idempotent, the right ideal eR is a projective right R-module. More is true: a right module M_R over a ring R is cyclic and projective if and only if $M_R \cong eR$ for some idempotent e of R.

Exercise 2.9.4
Show that if R is a ring, $e_1, \ldots, e_n$ are idempotents of R, $1 = e_1 + \cdots + e_n$ and $e_i e_j = 0$ for every $i, j = 1, \ldots, n$ with $i \neq j$, then $R_R = e_1 R \oplus \cdots \oplus e_n R$ and $_R R = R e_1 \oplus \cdots \oplus R e_n$.

Exercise 2.9.5
(a) Let M_R be a right module and $E := \text{End}(M_R)$ be its endomorphism ring. Show that there is a one-to-one correspondence φ between the set I of all idempotent elements of E and the set

$$\{ (A, B) \mid A, B \leq M_R, \ M_R = A \oplus B \}$$

of all pairs (A, B) of submodules of M_R whose sum is direct and equal to M_R. If $e \in I$, the corresponding pair is $\varphi(e) = (e(M_R), \ker e)$.

(b) This exercise allows us to give a different interpretations of Exercise 1.4.17, the exercise about non-unitary modules over a ring R with identity. Non-unitary modules ${}_R M$ can also be described as pair (M, λ), where M is an abelian group and $\lambda \colon R \to \mathrm{End}(M)$ is a mapping that respects addition and multiplication, but does not necessarily maps the identity 1_R of R to the identity ι_M of $\mathrm{End}(M)$. Nevertheless $e := \varphi(1)$ is an idempotent of $\mathrm{End}(M_R)$, as is easily seen, hence, by (a), it corresponds to a direct-sum decomposition $M = e(M) \oplus \ker(e)$ of the abelian group M. Show that $e(M) = M_1$ and $\ker(e) = M_0$ (notations as in Exercise 1.4.17).

Exercise 2.9.6
Show that if e is an idempotent of a ring R, then $eRe := \{ere \mid r \in R\}$ is a ring with respect to the operations induced by R. Its identity is e, so that if $e \neq 1_R$, then eRe is not a subring of R. Also show that the endomorphism ring of the right R-module eR is isomorphic to eRe. The isomorphism $\varphi \colon \mathrm{End}(eR) \to eRe$ is defined by $\varphi(f) = f(e)e$ for every $f \in \mathrm{End}(eR)$.

Exercise 2.9.7
Show that if R is a ring, I_λ is a non-zero right ideal of R for every $\lambda \in \Lambda$ and $R_R = \bigoplus_{\lambda \in \Lambda} I_\lambda$, then the index set Λ is finite.

Exercise 2.9.8
Show that if e is an idempotent of a ring R, then $eR \otimes_R M$ is canonically isomorphic to eM for every left R-module ${}_R M$, and $\mathrm{Hom}(eR, N_R)$ is canonically isomorphic to Ne for every right R-module N_R. More precisely:

(1) Show that the functors
$$eR \otimes_R - \colon R\text{-Mod} \to \mathbf{Ab} \quad \text{and} \quad e - \colon R\text{-Mod} \to \mathbf{Ab},$$
defined on objects by ${}_R M \mapsto eM$, are naturally isomorphic.

(2) Show that the functors
$$\mathrm{Hom}(eR, -) \colon \mathrm{Mod}\text{-}R \to \mathbf{Ab} \quad \text{and} \quad - e \colon \mathrm{Mod}\text{-}R \to \mathbf{Ab},$$
defined on objects by $N_R \mapsto Ne$, are naturally isomorphic.

[Similarly, the functors
$$- \otimes_R Re \colon \mathrm{Mod}\text{-}R \to \mathbf{Ab} \quad \text{and} \quad - e \colon \mathrm{Mod}\text{-}R \to \mathbf{Ab}$$
$(N_R \mapsto Ne)$ are naturally isomorphic.

The functors
$$\mathrm{Hom}(Re, -) \colon R\text{-Mod} \to \mathbf{Ab} \quad \text{and} \quad e - \colon R\text{-Mod} \to \mathbf{Ab} \ ({}_R M \mapsto eM),$$
are also naturally isomorphic.]

Exercise 2.9.9
Show that if R, S are rings, and $A_R, {}_R B_S, C_S$ are a right R-module, a bimodule, and a right S-module, respectively, then there is a canonical group isomorphism
$$\mathrm{Hom}_S(A \otimes_R B_S, C_S) \cong \mathrm{Hom}_R(A_R, \mathrm{Hom}_S({}_R B_S, C_S)_R).$$

[Hint: Both groups are canonically isomorphic to the subgroup of the group of all R-balanced mappings $\beta \colon A_R \times {}_R B \to C$ consisting of the R-balanced mappings such that $\beta(a, bs) = \beta(a, b)s$ for all $a \in A$, $b \in B$ and $s \in S$.]

Given a ring S, let proj-S denote the full subcategory of Mod-S whose objects are all finitely generated projective right S-modules.

If M_S is a right S-module, let add(M_S) denote the full subcategory of Mod-S whose objects are all modules isomorphic to direct summands of direct sums M^n of a finite number of copies of M. For example, proj-S = add(S_S).

Exercise 2.9.10
Let M_S be a right S-module and let $E = \operatorname{End}(M_S)$ be its endomorphism ring, so that ${}_E M_S$ is a bimodule. Consider the functors

$$\operatorname{Hom}_S(M, -) \colon \operatorname{Mod-}S \to \operatorname{Mod-}E \quad \text{and} \quad - \otimes_E M \colon \operatorname{Mod-}E \to \operatorname{Mod-}S.$$

Show that:
(a) The two functors map the object M_S to E_E and the object E_E to M_S respectively.
(b) The two functors map M_S^n to E_E^n and E_E^n to M_S^n respectively.
(c) The two functors respect direct summands, so that they map add(M_S) to proj-E and proj-E to add(M_S).
(d) The two functors induce an equivalence between the full subcategory add(M_S) of Mod-S and the full subcategory proj-E of Mod-E.
(e) If $e \in E$ is idempotent, the object eE of proj-E corresponds to the direct summand eM of M_S, which is an object of add(M_S).

Exercise 2.9.11
Let R be a ring with identity. We will denote by proj-R both the class of all finitely generated projective right R-modules and the full subcategory of Mod-R whose objects are all finitely generated projective right R-modules. In other words, we use the same symbol for the category proj-R and for its class of objects. For every ring R under consideration, we fix a set $V(R)$ of representatives up to isomorphism of the finitely generated projective right R-modules. Thus it is possible to associate to every module $A_R \in$ proj-R a unique module $\langle A_R \rangle \in V(R)$ isomorphic to A_R, and we get that, for all $A_R, B_R \in$ proj-R, $A_R \cong B_R$ if and only if $\langle A_R \rangle = \langle B_R \rangle$.
(a) Show that the set $V(R)$ is a commutative monoid with respect to the addition defined by $\langle A_R \rangle + \langle B_R \rangle := \langle A_R \oplus B_R \rangle$ for every $\langle A_R \rangle, \langle B_R \rangle \in V(R)$.

[The monoid $V(R)$ can also be viewed in an informal way as the monoid whose elements are the isomorphism classes of the modules in proj-R. More precisely, define, for every right R-module A_R, its *isomorphism class*

$$\langle A_R \rangle := \{ B_R \mid B_R \text{ is a module isomorphic to } A_R \},$$

and then set $V(R) = \{ \langle A_R \rangle \mid A_R \in \text{proj-}R \}$ and $\langle A_R \rangle + \langle B_R \rangle = \langle A_R \oplus B_R \rangle$ for all isomorphism classes $\langle A_R \rangle, \langle B_R \rangle \in V(R)$. But if we construct $V(R)$ in this way, the elements $\langle A_R \rangle$ of $V(R)$ are not sets, and thus $V(R)$ is not a set by Zermelo's Sum Axiom (Union Axiom) of General Set Theory ("for any set S there exists the set whose elements are the elements of the elements of S"). To avoid this set theoretical difficulty, we prefer not to introduce $V(R)$ as the class whose elements are the isomorphism classes of finitely generated projective modules, but to think of $V(R)$ as a fixed set of representatives of proj-R up to isomorphism.]

Any ring morphism $f: R \to S$ induces a monoid morphism

$$V(f): V(R) \to V(S),$$

defined by $V(f): \langle A_R \rangle \mapsto \langle A \otimes_R S \rangle$ for all $\langle A_R \rangle \in V(R)$.

(b) Show that $V(-)$ is a functor from the category **Rng** to the category **CMon** of commutative monoids.

[Clearly, the monoid $V(R)$ describes all direct sum decompositions of finitely generated projective right R-modules up to isomorphism, in the sense that to every decomposition of a projective module $A_R \in \text{proj-}R$ as a direct sum of finitely many submodules there corresponds a decomposition of the element $\langle A_R \rangle$ of the monoid $V(R)$ as a sum of elements of $V(R)$.]

(c) More generally, show that if M_S is any right S-module, it is possible to define a commutative monoid $V(\text{add}(M_S))$ whose elements are in one-to-one correspondence with the isomorphism classes $\langle N_R \rangle$ of the objects N_R of $\text{add}(M_S)$. The monoid $V(\text{add}(M_S))$ describes all direct-sum decompositions of the objects of $\text{add}(M_S)$. Notice that $V(S) = V(\text{add}(S_S))$.

Exercise 2.9.12

Let M_S be a right S-module and let $E := \text{End}(M_S)$ be its endomorphism ring. Show that the equivalence $\text{add}(M_S) \cong \text{proj-}E$ described in Exercise 2.9.10(d) induces a monoid isomorphism $V(\text{add}(M_S)) \cong V(E)$. Thus the direct-sum decompositions of a module M_S are described by the monoid $V(E)$ for a suitable ring E, that is, by finitely generated projective modules over a suitable ring.

Exercise 2.9.13

The functor $\text{Hom}(-, R): \text{Mod-}R \to R\text{-Mod}$ induces a duality

$$\text{proj-}R \to R\text{-proj},$$

i.e., a duality $\text{add}(R_R) \to \text{add}(_R R)$. Thus $V(\text{add}(R_R))$ is canonically isomorphic to $V(\text{add}(_R R))$. In other words, the monoid $V(R)$, associated to the ring R, does not depend on the fact that we have chosen finitely generated projective *right* modules. In order to define the monoid $V(R)$, we could have chosen finitely generated projective *left* modules and we would have obtained the same monoid.

2.10 Simple Modules, Semisimple Modules

A *simple* right module is a non-zero right module M_R whose submodules are only M_R and 0. Thus a simple module has exactly two submodules.

> **Lemma 2.10.1**
>
> A right module M_R is simple if and only it is isomorphic to R_R/I for some maximal right ideal I of R.

Proof Let M_R be a simple right module. Since $M_R \neq 0$, there exists a non-zero element $x \in M_R$. The mapping $\varphi: R_R \to M_R$ defined by $\varphi(r) = xr$ for every $r \in R$ is a right R-module homomorphism. Hence its kernel is a right ideal I of R,

2.10 Simple Modules, Semisimple Modules

and its image is a submodule of M_R that contains the non-zero element x. As M_R is simple, φ must be surjective. By the Correspondence Theorem 1.9.3, there is a one-to-one correspondence between the set of all right ideals of R containing $\ker \varphi$ and the set of all submodules of M_R. As M_R is simple, M_R has two submodules, hence there exist exactly two right ideals of R containing I, which are necessarily I itself and R. Hence the right ideal I is maximal. By the First Isomorphism Theorem, $M_R \cong R_R/I$. □

Lemma 2.10.2 (Schur's Lemma)
The endomorphism ring of a simple module is a division ring.

Proof To see it, take $\varphi \colon M_R \to M_R$, a non-zero homomorphism. φ is surjective because $\varphi(M)$ is a non-zero submodule of M, so it is M, and φ is injective because $\ker(\varphi)$ is a proper submodule of M, so it is 0. □

We record a consequence that will be useful in the sequel.

Corollary 2.10.3
Let S_R, S'_R be simple right R-modules

(i) *If S_R, S'_R are not isomorphic, then $\mathrm{Hom}(S_R, S'_R) = 0$*
(ii) *If S_R, S'_R are isomorphic and $D := \mathrm{End}(S_R)$, then $\mathrm{Hom}(S_R, S'_R)$ is a one-dimensional right vector space over the division ring D.*

Proof
(i) If S_R, S'_R are simple modules and $\varphi \colon S_R \to S'_R$ is a non-zero homomorphism, then the kernel of φ must be a proper submodule of S_R. As S_R is simple, it follows that $\ker \varphi = 0$. Thus φ is injective, it follows that S_R is isomorphic to a submodule of S'_R. But $S_R \neq 0$ and S'_R is simple, so that S_R and S'_R must be isomorphic.
(ii) We know that $_D S_R$ is a bimodule, so that we have the functor

$$\mathrm{Hom}(_D S_R, -) \colon \mathrm{Mod}\text{-}R \to \mathrm{Mod}\text{-}D = \mathbf{Vect}\text{-}D.$$

Since functors sends isomorphisms to isomorphisms, $S_R \cong S'_R$ implies that the right D-vector spaces $\mathrm{Hom}(_D S_R, S_R)$ and $\mathrm{Hom}(_D S_R, S'_R)$ are isomorphic, that is $D_D \cong \mathrm{Hom}(_D S_R, S'_R)$. □

A module M_R is *semisimple* if every submodule of M_R is a direct summand of M_R.

For instance:

(1) Every simple module is semisimple.
(2) If R is a division ring, every module over R is semisimple.

Notice that a module M_R is semisimple if and only if every short exact sequence with M_R in the middle, that is, every short exact sequence of the form $0 \to A_R \to M_R \to C_R \to 0$, splits.

Lemma 2.10.4
Submodules and homomorphic images of semisimple modules are semisimple modules.

Proof Let M_R be semisimple. Fix $N \leq M_R$ and a submodule $P \leq N$. We must show that P, which is a direct summand of M, is also a direct summand of N. Since P is a direct summand of M, we have that $M = P \oplus C$ for some submodule C of M. From Exercise 1.8.1, we know that $N = P \oplus (C \cap N)$. Therefore N is semisimple. In order to show that M/N is semisimple, notice that N is a direct summand of M, that is, $M = N \oplus D$ for some submodule D of M, so $M/N \cong D$, and D is semisimple, because, as we have just seen, submodules of semisimple modules are semisimple. □

In Theorem 2.10.7 we will give some characterizations of semisimple modules. To prove it, we need the following lemma.

Lemma 2.10.5
Let M_R be a module that is a sum of simple submodules, that is, $M_R = \sum_{\lambda \in \Lambda} S_\lambda$, where each S_λ is a simple submodule of M_R. Then for any $N \leq M_R$ there exists a subset X of Λ such that $M_R = N \oplus (\oplus_{\lambda \in X} S_\lambda)$.

Proof Set

$$\mathcal{F} := \{ Y \subseteq \Lambda \mid N + \sum_{\lambda \in Y} S_\lambda = N \oplus (\oplus_{\lambda \in Y} S_\lambda) \},$$

that is, $\mathcal{F}$ is the set of all subsets Y of Λ for which the sum $N + \sum_{\lambda \in Y} S_\lambda$ is direct. This set $\mathcal{F}$ is non-empty, because the empty set belongs to $\mathcal{F}$. Partial order $\mathcal{F}$ by set inclusion. We want to apply Zorn's Lemma to this partially ordered set.

2.10 Simple Modules, Semisimple Modules

Let $\{Y_i \mid i \in I\}$ be a chain in $\mathcal{F}$, and assume that $\bigcup_{i \in I} Y_i \notin \mathcal{F}$. Then the sum $N + (\sum_{\lambda \in \bigcup_i Y_i} S_\lambda)$ is not a direct sum, so that there exist a positive integer m, distinct elements $\lambda_1, \ldots, \lambda_m$ in $\bigcup_{i \in I} Y_i$, an element $n \in N$ and elements $s_{\lambda_1} \in S_{\lambda_1}, \ldots, s_{\lambda_m} \in S_{\lambda_m}$ not all zero, such that $n + s_{\lambda_1} + \cdots + s_{\lambda_m} = 0$. Now $\lambda_1, \ldots, \lambda_m \in Y_{i_0}$ for some index $i_0 \in I$. Thus the sum $N + \sum_{\lambda \in Y_{i_0}} S_\lambda$ is not direct, which is a contradiction to $Y_{i_0} \in \mathcal{F}$. Hence we can apply Zorn's Lemma, and we find that there exists a maximal element X in $\mathcal{F}$. To prove the Lemma, it suffices to show that $M = N \oplus (\oplus_{\lambda \in X} S_\lambda)$. We already know that the sum $N + \sum_{\lambda \in X} S_\lambda$ is direct, and clearly $M \supseteq N + \sum_{\lambda \in X} S_\lambda$. It remains only to show that $M \subseteq N + \sum_{\lambda \in X} S_\lambda$. Equivalently, that $S_{\bar\lambda} \subseteq N + \sum_{\lambda \in X} S_\lambda$ for every $\bar\lambda \in \Lambda$. Assume the contrary, so that $S_{\bar\lambda} \not\subseteq N + \sum_{\lambda \in X} S_\lambda$ for some $\bar\lambda \in \Lambda$. Then $S_{\bar\lambda} \cap (N + \sum_{\lambda \in X} S_\lambda)$ is a submodule of $S_{\bar\lambda}$ different from $S_{\bar\lambda}$, so that $S_{\bar\lambda} \cap (N + \sum_{\lambda \in X} S_\lambda) = 0$. It follows that the sum

$$S_{\bar\lambda} \oplus (N + \sum_{\lambda \in X} S_\lambda) = S_{\bar\lambda} \oplus (N \oplus (\oplus_{\lambda \in Y} S_\lambda)) = N \oplus \sum_{\lambda \in X \cup \{\bar\lambda\}} S_\lambda$$

is direct. Thus $X \cup \{\bar\lambda\} \in \mathcal{F}$. This contradicts the maximality of X. □

Definition 2.10.6

Let M_R be a right R-module. The *socle* $\mathrm{soc}(M_R)$ of M_R is the sum of all simple submodules of M_R.

Thus $\mathrm{soc}(M) = 0$ if and only if M has no simple submodules.

Theorem 2.10.7
The following conditions are equivalent for a right R-module M:

(i) *M is a sum of simple submodules, that is, M is equal to its socle.*
(ii) *M is a direct sum of simple submodules.*
(iii) *M is semisimple.*

Proof
(i) $\Rightarrow$ (ii) By Lemma 2.10.5 (applied to the case $N = 0$), we have that any sum $\sum_{\lambda \in \Lambda} S_\lambda$ of simple modules is a direct sum of simple modules, i.e., is equal to $\sum_{\lambda \in X} S_\lambda$ for some subset X of Λ.

(ii) $\Rightarrow$ (iii) By Lemma 2.10.5, any submodule of a sum M_R of simple modules is a direct summand of M_R, that is, any (direct) sum of simple modules is a semisimple module.

(iii) ⇒ (i) Let M be a semisimple module and assume that $M \neq \mathrm{soc}(M)$. Let x be an element of M not in $\mathrm{soc}(M)$. Set

$$\mathcal{F} := \{ N \mid N \leq M,\ N \supseteq \mathrm{soc}(M),\ x \notin N \}.$$

The set $\mathcal{F}$ is non-empty, for instance $\mathrm{soc}(M) \in \mathcal{F}$. Partially order $\mathcal{F}$ by set inclusion. It is easy to see that it is possible to apply Zorn's Lemma. By Zorn's Lemma, the partially ordered set $\mathcal{F}$ has a maximal element N_0. Since M is semisimple, we have $M = N_0 \oplus C$ for some submodule C of M. In particular, C is semisimple. From $x \notin N_0$, it follows that N_0 is a proper submodule of M, so that $C \neq 0$. Now we have two cases. If C is a simple module, then C is contained in $\mathrm{soc}(M) \subseteq N_0$, a contradiction because $M = N_0 \oplus C$ and $C \neq 0$. If C is not a simple module, C has a proper non-zero submodule A. As C is semisimple, we have that $C = A \oplus B$ for some proper non-zero submodule B of C. Thus both $N_0 \oplus A$ and $N_0 \oplus B$ properly contain N_0. From the maximality of N_0 in $\mathcal{F}$, it follows that $N_0 \oplus A$ and $N_0 \oplus B$ do not belong to $\mathcal{F}$. Hence x belongs to both $N_0 \oplus A$ and $N_0 \oplus B$, so that $x \in (N_0 \oplus A) \cap (N_0 \oplus B) = N_0$, another contradiction. This shows that $M = \mathrm{soc}(M)$. □

Thus every semisimple module is a direct sum of simple submodules. Let us pass to consider the "uniqueness up to isomorphism" of such a direct-sum decomposition.

Proposition 2.10.8
If $M_R = \oplus_{i \in I} S_i = \oplus_{j \in J} S'_j$ with the S_i, S'_j simple submodules of M_R, then there exists a bijection $\varphi \colon I \to J$ such that $S_i \cong S'_{\varphi(i)}$ for every $i \in I$.

Proof We claim that for every simple module S_R, the sets $I(S_R) := \{ i \in I \mid S_i \cong S_R \}$ and $J(S_R) := \{ j \in J \mid S'_j \cong S_R \}$ have the same cardinality. If this claim is true, then for every simple module S_R we can fix a bijection $\varphi_{S_R} \colon I(S_R) \to J(S_R)$. The (non-empty) sets $I(S_R)$ and $J(S_R)$ form a partition of I and J, respectively, so that, gluing together these functions $\varphi_{S_R} \colon I(S_R) \to I(S_R)$, we get the bijection $\varphi \colon I \to J$ with the required property. This proves the Proposition.

In order to prove the claim, apply the functor

$$\mathrm{Hom}({}_D S_R, -) \colon \mathrm{Mod}\text{-}R \to \mathrm{Mod}\text{-}D,$$

where D is the division ring $\mathrm{End}(S_R)$, to the module $M_R = \oplus_{i \in I} S_i$. Then

$$\mathrm{Hom}(S_R, M_R) = \mathrm{Hom}(S_R, \oplus_{i \in I} S_i) \cong \oplus_{i \in I} \mathrm{Hom}(S_R, S_i),$$

because the module S_R is finitely generated. But $\mathrm{Hom}(S_R, S_i)$ is 0 if $S_i \not\cong S$, and is $\cong D_D$ if $S_i \cong S$ (Corollary 2.10.3). So

$$\dim_D(\,\mathrm{Hom}(S_R, M_R)) = |\{\,i \in I \mid S_i \cong S\,\}|.$$

Thus $|\{\,i \in I \mid S_i \cong S\,\}|$ does not dipend on the direct-sum decomposition of M_R into simple submodules. □

2.11 Noetherian Modules, Artinian Modules

Proposition 2.11.1
The following conditions are equivalent for a partially ordered set $(P, \le)$:

(i) *Every non-empty subset of $(P, \le)$ has a maximal element (that is, for every $X \subseteq P$ with $X \ne \emptyset$, there exists a maximal element in the set X partially ordered by the order induced by the order $\le$ of P.)*
(ii) *Every countable ascending chain in P is stationary (that is, if $p_0 \le p_1 \le p_2 \le \ldots$ are elements of P, then there exists $n \ge 0$ with $p_n = p_{n+1} = p_{n+2} = \ldots$).*
(iii) *Every non-empty chain C in P has a greatest element.*

Proof
(i) $\Rightarrow$ (iii). follows from the fact that in a chain an element is maximal if and only if it is the greatest element.
(iii) $\Rightarrow$ (ii). Let $p_0 \le p_1 \le p_2 \le \ldots$ be a countable ascending chain. The chain $\mathcal{C} = \{\,p_i \mid i \ge 0\,\}$ has a greatest element by (iii). Assume p_n is the greatest element of $\mathcal{C}$. Then $p_n = p_{n+1} = p_{n+2} = \ldots$
(ii) $\Rightarrow$ (i). We will show that if P is a partially ordered set for which (i) does not hold, then (ii) also does not hold. Let X be a non-empty subset of $(P, \le)$ that does not have maximal elements. Construct by induction a sequence p_n of elements of X in the following way. For $n = 0$, fix any p_0 in X (such a p_0 exists because X is non-empty). Assume that $p_n \in X$ has been defined. Then p_n is not maximal in X, so that there is a $p_{n+1} \in X$ with $p_{n+1} > p_n$. Then (ii) does not hold for the ascending chain $p_0 \le p_1 \le p_2 \le \ldots$
□

A *noetherian* partially ordered set is a partially ordered set which satisfies the equivalent conditions of the above proposition. An *artinian* partially ordered set is a partially ordered set $(P, \le)$ such that its opposite $(P, \ge)$ is noetherian.

A module M_R is *noetherian* if the lattice $\mathcal{L}(M_R)$ of all the submodules of M_R is a noetherian partially ordered set with respect to inclusion. A module M_R is *artinian* if the lattice $\mathcal{L}(M_R)$ is an artinian partially ordered set with respect to inclusion. Thus M_R is noetherian if every ascending chain $N_0 \leq N_1 \leq N_2 \leq \ldots$ of submodules of M_R is stationary, and is artinian if every descending chain $N_0 \geq N_1 \geq N_2 \geq \ldots$ of submodules of M_R is stationary.

A ring R is *right noetherian (left noetherian)* if the right R-module R_R is noetherian (the left R-module $_RR$ is noetherian). A ring R is *right artinian (left artinian)* if R_R ($_RR$) is artinian.

Examples 2.11.2
(1) Simple modules are both noetherian and artinian. Finite modules, that is, modules with finitely many elements are noetherian and artinian.
(2) The abelian group $\mathbb{Z}(p^\infty)$, where p is a prime, is a $\mathbb{Z}$-module that is artinian, but not noetherian.
(3) The $\mathbb{Z}$-module $\mathbb{Q}$ is neither noetherian nor artinian. In fact, $\{2^n\mathbb{Z} \mid n \geq 0\}$ is a strictly descending chain of subgroups of $\mathbb{Q}$ that is not stationary, and $\{\frac{1}{2^n}\mathbb{Z} \mid n \geq 0\}$ is a strictly ascending chain of subgroups of $\mathbb{Q}$ that is not stationary.
(4) The abelian group $\mathbb{Z}$ is a $\mathbb{Z}$-module that is noetherian, but not artinian. In fact, the submodules of $\mathbb{Z}_\mathbb{Z}$ are the subgroups $n\mathbb{Z}$ with n a non-negative integer. Moreover, $n\mathbb{Z} \subseteq m\mathbb{Z}$ if and only if $m|n$ (m divides n). The module $\mathbb{Z}_\mathbb{Z}$ is not artinian, because the chain

$$\mathbb{Z} \supset 2\mathbb{Z} \supset 4\mathbb{Z} \supset \cdots \supset 2^k\mathbb{Z} \supset \ldots$$

is not stationary. The module $\mathbb{Z}_\mathbb{Z}$ is noetherian, because if

$$n_0\mathbb{Z} \subseteq n_1\mathbb{Z} \subseteq \ldots$$

is an ascending chain of submodules, then $n_1|n_0, n_2|n_1, n_3|n_2, \ldots$, that is, in the sequence $n_0, n_1, n_2, \ldots$, every n_i divides the previous element n_{i-1}. Hence, we have two cases. If $n_i = 0$ for every i, the chain is stationary. If $n_i \neq 0$ for some i, then $n_i \geq n_{i+1} \geq n_{i+2} \geq \ldots$ is a sequence of non-negative integers, hence must be stationary as well.

The fact that $\mathbb{Z}_\mathbb{Z}$ is noetherian also follows from the next Proposition.

Proposition 2.11.3
A module A_R is noetherian if and only if every submodule of A_R is finitely generated.

Proof Assume A_R is noetherian. Let B be a submodule of A_R. Let $\mathcal{F}$ be the set of all finitely generated submodules of B. The set $\mathcal{F}$ is non-empty, because it contains the zero module. Since A_R is noetherian, $\mathcal{F}$ must have a maximal element $\overline{B}$. Let us prove that $\overline{B} = B$. Otherwise, $\overline{B}$ would be a proper submodule of B, so that if we fix an element $x_0 \in B \setminus \overline{B}$, we would have $\overline{B} \subset \overline{B} + x_0R \in \mathcal{F}$. This contradicts the maximality of $\overline{B}$. Thus $\overline{B} = B$, so that $B \in \mathcal{F}$. That is, B is finitely generated.

2.11 Noetherian Modules, Artinian Modules

Conversely, assume that all submodules of A_R are finitely generated. Let

$$A_0 \subseteq A_1 \subseteq A_2 \subseteq \ldots$$

be an ascending chain of submodules of A_R. Then $\bigcup_{i \geq 0} A_i$ is a submodule of A_R. Hence $\bigcup_{i \geq 0} A_i$ is finitely generated. Let $x_1, \ldots, x_t$ be generators of this module $\bigcup_{i \geq 0} A_i$. For every $j = 1, \ldots, t$, the element x_j belongs to some $A_{n(j)}$. If n_0 is greater than $n(1), \ldots, n(t)$, then $x_j \in A_{n_0}$ for every $j = 1, \ldots, n$, so that $\bigcup_{i \geq 0} A_i \subseteq A_{n_0}$. It follows that $A_i = A_{n_0}$ for every $i \geq n_0$. Hence the chain is stationary. □

Proposition 2.11.4
Let

$$0 \to A_R \to B_R \to C_R \to 0$$

be a short exact sequence of right R-modules. Then

(i) B_R is noetherian if and only if A_R and C_R are noetherian.
(ii) B_R is artinian if and only if A_R and C_R are artinian.

Proof We prove (i) only. The proof for (ii) is similar. Without loss of generality assume that A_R is a submodule of B_R and $C_R = B_R/A$, that is, we can assume that the short exact sequence is the canonical short exact sequence $0 \to A_R \hookrightarrow B_R \to B_R/A \to 0$.

Assume B_R noetherian. Then A_R is obviously noetherian, because every ascending chain of submodules of A is an ascending chain of submodules of B. If

$$C_0 \subseteq C_1 \subseteq C_2 \subseteq \ldots$$

is an ascending chain of submodules of B_R/A, then $C_i = B_i/A$ for some submodule B_i of B_R containing A_R. Then

$$B_0 \subseteq B_1 \subseteq B_2 \subseteq \ldots$$

is an ascending chain of submodules of B, hence it is stationary, so that the chain of the C_i's is stationary as well. This proves that B_R/A also is noetherian.

For the converse, assume A and B/A noetherian modules. We must prove that B is noetherian. Let

$$B_0 \subseteq B_1 \subseteq B_2 \subseteq \ldots$$

be an ascending chain of submodules of B. Then

$$A \cap B_0 \subseteq A \cap B_1 \subseteq A \cap B_2 \subseteq \ldots$$

and

$$(B_0 + A)/A \subseteq (B_1 + A)/A \subseteq (B_2 + A)/A \subseteq \ldots$$

are ascending chains of submodules of A and B/A respectively, hence they are stationary. Thus there exists $n_0 \geq 0$ such that $A \cap B_i = A \cap B_{n_0}$ and $B_i + A = B_{n_0} + A$ for every $i \geq n_0$. Let us prove that $B_i = B_{n_0}$ for every $i \geq n_0$.

Clearly, $B_i \supseteq B_{n_0}$ because the chain of the B_i's is ascending.

Conversely, let b be an element in B_i. Then $b + A \in B_i + A = B_{n_0} + A$, so $b = b' + a$ for suitable $b' \in B_{n_0}$ and $a \in A$. Then $b - b' = a \in A$, so that $b - b' \in B_i \cap A = B_{n_0} \cap A$. Thus $b - b' \in B_{n_0}$, and this implies that $b \in B_{n_0}$, because $b' \in B_{n_0}$. □

Corollary 2.11.5
Let $A_1, \ldots, A_n$ be right R-modules. Then $A_1 \oplus A_2 \oplus \cdots \oplus A_n$ is a noetherian (artinian) module if and only if A_i is a noetherian (artinian) module for every $i = 1, \ldots, n$.

Proof If $A_1 \oplus A_2 \oplus \cdots \oplus A_n$ is noetherian, then every A_i is noetherian, because submodules of noetherian modules are noetherian modules. For the converse, we argue by induction of $n \geq 1$ For $n = 1$ the statement is trivial. Assume that the statement is true for $n - 1$. Let $A_1, \ldots, A_n$ be noetherian modules. Then $A_1 \oplus \cdots \oplus A_{n-1}$ is noetherian by the inductive hypothesis, and $(A_1 \oplus \cdots \oplus A_n)/(A_1 \oplus \cdots \oplus A_{n-1}) \cong A_n$ is noetherian. Thus we can apply the previous Proposition. Similarly for artinian modules. □

Corollary 2.11.6
If R is a right noetherian (right artinian) ring and M_R is a finitely generated right R-module, then M_R is a noetherian (artinian) module.

Proof We give the proof for the case of a right artinian ring. Let R be a right artinian ring, that is, suppose that the module R_R is artinian. By the previous Corollary, R_R^n is an artinian module for every $n \geq 0$. The finitely generated module M_R is a homomorphic image of R_R^n for some n. Therefore M_R is artinian by Proposition 2.11.4(ii). □

2.12 Modules Over the Factor Ring R/I

Exercise 2.11.7
Set
$$R := \left\{ \begin{pmatrix} q & 0 \\ r & s \end{pmatrix} \Big| q \in \mathbb{Q},\ r, s \in \mathbb{R} \right\}.$$

The ring R is usually denoted as $\begin{pmatrix} \mathbb{Q} & 0 \\ \mathbb{R} & \mathbb{R} \end{pmatrix}$. A similar notation is used whenever one has to denote the rings defined like R, that is, the subrings of a ring $\mathbb{M}_n(S)$ whose entries belong to subgroups of S.

(a) Show that R is left artinian and left noetherian, but it is neither right artinian nor right noetherian.

(b) Show that the ring $\begin{pmatrix} \mathbb{Q} & 0 \\ \mathbb{Q} & \mathbb{Z} \end{pmatrix}$ is right noetherian but not left noetherian, and is neither left nor right artinian.

(c) Show that the ring $\begin{pmatrix} \mathbb{Z} & 0 \\ \mathbb{Q} & \mathbb{Q} \end{pmatrix}$ is left noetherian but not right noetherian, and is neither left nor right artinian. The structure of the right (left) ideals of this ring will be also studied in Exercise 3.1.6.

(d) Show that the ring $\begin{pmatrix} \mathbb{Q} & 0 \\ \mathbb{Q}(x) & \mathbb{Q}(x) \end{pmatrix}$ is left artinian but not right artinian. Here $\mathbb{Q}(x)$ is the field of fractions of the polynomial ring $\mathbb{Q}[x]$.

2.12 Modules Over the Factor Ring R/I

Let R be a ring and $I \trianglelefteq R$ be a two-sided ideal of R. What are R/I-modules? If M_R is a right R-module and $MI = 0$, then M becomes a right R/I-module setting $m(r + I) := mr$. This operation is well defined because $MI = 0$. Conversely, if $M_{R/I}$ is a R/I-right module, then M becomes a right R-module via $mr := m(r + I)$. Equivalently, if $M_{R/I}$ is a right R/I-module, we have a canonical antihomomorphism $R/I \to \mathrm{End}_\mathbb{Z}(M)$. Compose it with $R \to R/I$ and get an antihomomorphism $R \to \mathrm{End}_\mathbb{Z}(M)$, i.e., a right R-module M_R. Note that $MI = 0$.

So right R-modules M with $MI = 0$ essentially coincide with right R/I-modules.

Using the categorical language, we can be more precise:

> **Lemma 2.12.1**
> *Let R be a ring and $I \trianglelefteq R$. The category Mod-R/I is isomorphic to the full subcategory of Mod-R whose objects are all right R-modules M such that $MI = 0$.*

Proof Let $\mathcal{A}_I$ be the full subcategory of Mod-R whose objects are all right R-modules M such that $MI = 0$. Let

$$F: \text{Mod-}R/I \to \mathcal{A}_I$$

be defined as follows. Associate to $M_{R/I}$ the abelian group M with the R-module structure defined by $x \cdot r := x(r + I)$ for every $x \in M$, $r \in R$. If $i \in I$, then $x \cdot i = 0$, so that $MI = 0$. So $M_R \in \text{Ob}(\mathcal{A}_I)$. If $f: M_{R/I} \to M'_{R/I}$ is a morphism in Mod-R/I, f is also a morphism in Mod-R, so that f can be mapped to f via F.

Let $G: \mathcal{A}_I \to \text{Mod-}R/I$ be the functor defined as follows: if $M_R \in \text{Ob}(\mathcal{A}_I)$, then M is a right R-module and $MI = 0$. The abelian group M becomes a right R/I-module via $x \cdot (r + I) = xr$. This right scalar multiplication is well defined, because if $r + I = r' + I$, then $r - r' \in I$ and $x(r - r') = 0$, so $xr = xr'$. If $f: M_R \to M'_R$ is a morphism of R-modules, $MI = 0$ and $M'I = 0$, then f is also an R/I-module morphism, so that f can be mapped to f via G.

It is now easily seen that

$$G \circ F = 1_{\text{Mod-}R/I} \quad \text{and} \quad F \circ G = 1_{\mathcal{A}_I},$$

as we wanted to prove. □

Notice that if M, N are right R-modules with $MI = 0$ and $NI = 0$, so that M and N are right R/I-modules, the morphisms of M into N in the category Mod-R coincide with those in the category Mod-R/I, that is, a mapping $f: M \to N$ is a right R-module morphism if and only if it is a right R/I-module morphism.

Similarly, notice that if M_R is an R-module with $MI = 0$, a subgroup A of the additive group of M is a submodule of $M_{R/I}$ if and only if it is a submodule of M_R. Thus M_R and $M_{R/I}$ have the same lattice of submodules, that is, $\mathcal{L}(M_R) = \mathcal{L}(M_{R/I})$. From this elementary fact, we get most of the following Proposition:

Proposition 2.12.2
Let M_R be a module and let I be a two-sided ideal of R with $MI = 0$. Then

(i) *M_R is simple if and only if $M_{R/I}$ is simple.*
(ii) *M_R is semisimple if and only if $M_{R/I}$ is semisimple.*
(iii) *M_R is artinian if and only if $M_{R/I}$ is artinian.*
(iv) *M_R is noetherian if and only if $M_{R/I}$ is noetherian.*
(v) *M_R is finitely generated if and only if $M_{R/I}$ is finitely generated.*

Proof For (v), it suffices to notice that $x_1, x_2, \ldots, x_n$ generate M_R if and only if they generate $M_{R/I}$. □

> **Corollary 2.12.3**
> *If R is a right artinian ring (right noetherian ring) and I is a two-sided ideal of R, then R/I is a right artinian ring (right noetherian ring, respectively).*

Proof Let R be a right artinian ring. Then R_R is an artinian R-module. Hence R_R/I is an artinian R-module by Proposition 2.11.4(ii), and $(R_R/I)I = 0$. By Proposition 2.12.2(iii), the right R/I-module $(R/I)_{R/I}$ is artinian, that is, R/I is a right artinian ring. □

A subring of a right artinian (right noetherian) ring is not necessarily right artinian (right noetherian). For instance, consider the ring $R = k[X]$, where k is a field and X is an infinite set of indeterminates. Let K be the field of fractions of R. Then K is both artinian and noetherian, but its subring R is neither artinian nor noetherian.

2.13 Series of Modules: Modules of Finite Composition Length

Let M_R be a module. A *series* for M_R is a finite chain

$$0 = M_0 \subseteq M_1 \subseteq M_2 \subseteq \cdots \subseteq M_n = M_R \qquad (2.4)$$

of submodules of M_R. The *factors* of the series are the modules M_i/M_{i-1} for $i = 1, \ldots, n$. The *length* of the series is n. The series (2.4) is called a *composition series* for M_R if, for every $i = 1, \ldots, n$, the factor M_i/M_{i-1} is a simple module (equivalently, M_{i-1} is a maximal submodules of M_i). Two series

$$0 = M_0 \subseteq M_1 \subseteq \cdots \subseteq M_n = M_R \qquad (2.5)$$

and

$$0 = M'_0 \subseteq M'_1 \subseteq \cdots \subseteq M'_m = M_R \qquad (2.6)$$

of M_R are *equivalent* if $n = m$ and there exists a permutation σ of $\{1, 2, \ldots, n\}$ such that $M_i/M_{i-1} \cong M_{\sigma(i)}/M_{\sigma(i)-1}$ for every $i = 1, \ldots, n$. That is, two series are equivalent if and only if they have the same length and the same factors up to the order and isomorphism.

The series (2.5) is a *refinement* of the series (2.6) if there exist integers $0 = t_0 < t_1 < t_2 < \cdots < t_m = n$ such that $M'_i = M_{t_i}$ for every $i = 0, 1, 2, \ldots, m$. That is, a refinement is obtained by inserting more submodules in the series. If, moreover, $\{M_0, M_1, \ldots, M_n\} \supset \{M'_0, M'_1, \ldots, M'_m\}$, series (2.5) is a *proper refinement* of series (2.6).

Thus a series

$$0 = M_0 \subset M_1 \subset \cdots \subset M_n = M_R \qquad (2.7)$$

is a composition series if and only if it is a series without proper refinements, that is, all the refinements of (2.7) are obtained from (2.7) inserting submodules that are already in (2.7).

Examples 2.13.1
(1) Simple modules have a composition series.
(2) For a fixed prime p, the Prüfer group $\mathbb{Z}(p^\infty)$ has no composition series. To see it, recall that the partially ordered set $\mathcal{L}(\mathbb{Z}(p^\infty))$ is isomorphic to the well-ordered set $\{0, 1, 2, 3, \ldots, +\infty\}$ with its natural order, so that the last factor of any series of $\mathbb{Z}(p^\infty)$ cannot be simple.
(3) It is easily seen (exercise for the reader) that an abelian group has a composition series if and only if it is finite.

For the proof of the following theorem, we essentially follow Baumslag [3].

Theorem 2.13.2 (Schreier-Zassenhaus)
Any two series of a module have equivalent refinements.

Proof Suppose that (2.5) and (2.6) are two series of a module M_R.
For each i, we have the series

$$0 = M_i \cap M_0' \subseteq M_i \cap M_1' \subseteq \cdots \subseteq M_i \cap M_m' = M_i$$

for M_i, so that

$$M_{i-1} = M_{i-1} + 0 = M_{i-1} + (M_i \cap M_0') \subseteq M_{i-1} + (M_i \cap M_1') \subseteq \cdots \subseteq M_{i-1} + (M_i \cap M_m') = M_{i-1} + M_i = M_i$$

is an ascending chain of length m of submodules from M_{i-1} to M_i, for each $i = 1, 2, \ldots, n$. Gluing together all these n chains, we get a series for the module M_R of length mn, which is clearly a refinement of the series (2.5). The factors of this refinement are the mn modules $(M_{i-1} + (M_i \cap M_j'))/(M_{i-1} + (M_i \cap M_{j-1}'))$, for $i = 1, 2, \ldots, n$ and $j = 1, 2, \ldots, m$.

Inverting the roles of the series (2.5) and (2.6), one similarly gets a series for M_R, which is a refinement of the series (2.6), is of length nm, and whose factors are the nm modules $(M_{j-1}' + (M_j' \cap M_i))/(M_{j-1}' + (M_j' \cap M_{i-1}))$, for $j = 1, 2, \ldots, m$

2.13 Series of Modules: Modules of Finite Composition Length

and $i = 1, 2, \ldots, n$. In order to conclude that the two refinements are equivalent, it suffices to show that

$$(M_{i-1} + (M_i \cap M'_j))/(M_{i-1} + (M_i \cap M'_{j-1}))$$
$$\cong (M'_{j-1} + (M'_j \cap M_i))/(M'_{j-1} + (M'_j \cap M'_{i-1})),$$

for every $i = 1, 2, \ldots, n$ and every $j = 1, 2, \ldots, m$. To this end, it is convenient to prove the following lemma. □

Lemma 2.13.3
Let A, B, C be submodules of a right R-module D_R and suppose $B \le A$. Then the R-modules $A/(B + (A \cap C))$ and $(A + C)/(B + C)$ are isomorphic.

Proof Consider the canonical epimorphism $\pi : A \to (A + C)/(B + C)$, defined by $\pi(a) = a + (B + C)$ for every $a \in A$. An element $a \in A$ is in the kernel of π if and only if $a \in B + C$, that is, if and only if $a \in A \cap (B + C)$. Since $B \le A$, we have that $B + (A \cap C) = A \cap (B + C)$. Therefore $\ker \pi = B + (A \cap C)$. This allows us to conclude the proof of the Lemma. □

In order to conclude the proof of the Schreier-Zassenhaus Theorem, apply the Lemma to $A = M_i \cap M'_j$, $B = M_i \cap M'_{j-1}$ and $C = M_{i-1}$. We find that

$$(M_i \cap M'_j)/((M_i \cap M'_{j-1}) + (M_i \cap M'_j \cap M_{i-1}))$$
$$\cong ((M_i \cap M'_j) + M_{i-1})/((M_i \cap M'_{j-1}) + M_{i-1}).$$

This is the factor of the first refinement corresponding to the indices i and j. Similarly, inverting the roles of the series (2.5) and (2.6), we have from the Lemma that $(M'_j \cap M_i)/((M'_j \cap M_{i-1}) + (M'_j \cap M_i \cap M'_{j-1})) \cong ((M'_j \cap M_i) + M'_{j-1})/((M'_j \cap M_{i-1}) + M'_{j-1})$, the factor of the second refinement corresponding to the indices i and j. We now easily get that the two refinements are equivalent. This concludes the proof of the Theorem. □

Theorem 2.13.4 (Jordan-Hölder)
If a module M_R has a composition series of length n, then all its composition series are equivalent, and any series $0 = M_0 \subset M_1 \subset M_2 \subset \cdots \subset M_m = M_R$ of M_R can be refined to a composition series. In particular, any properly ascending series of submodules of M_R has length at most n.

Proof Let

$$0 = M_0 \subseteq M_1 \subseteq M_2 \subseteq \cdots \subseteq M_n = M, \tag{2.8}$$

$$0 = M'_0 \subseteq M'_1 \subseteq M'_2 \subseteq \cdots \subseteq M'_m = M \tag{2.9}$$

be two composition series of M_R. We must show that they are equivalent. By the Schreier-Zassenhaus Theorem, (2.8) and (2.9) have equivalent refinements. But composition series do not have proper refinements. Hence the factors of a refinement of a composition series consist of the factors of the composition series, which are simple modules, plus some copies of the zero module. Thus the non-zero factors of the two refinements are isomorphic, up to the order, that is, up to a permutation. Thus the factors of (2.8) and (2.9) are isomorphic, up to a permutation. Therefore the composition series (2.8) and (2.9) are equivalent.

If (2.8) is a composition series and (2.9) is an arbitrary series in which all inclusions are proper, then the Schreier-Zassenhaus Theorem yields two equivalent refinements

$$0 = N_0 \subseteq N_1 \subseteq N_2 \subseteq \cdots \subseteq N_t = M, \tag{2.10}$$

$$0 = N'_0 \subseteq N'_1 \subseteq N'_2 \subseteq \cdots \subseteq N'_t = M \tag{2.11}$$

of (2.8) and (2.9), respectively. Here $t \geq n$ and $t \geq m$. As before, (2.8) does not have proper refinements. Hence the factors N_i/N_{i-1} of the refinement (2.10) of a composition series (2.8) consist of the n factors of the composition series (2.8), which are simple modules, plus $t - n$ copies of the zero module. Hence the factors of (2.11) consist of the n simple factors of the composition series (2.8) plus $t - n$ copies of the zero module. Canceling the repetitions in (2.11) we get a series of length n which is still a refinement of (2.9) (and so $n \geq m$), and whose factors are all simple modules. Hence this new refinement of (2.9) is a composition series. □

By the Jordan-Hölder Theorem, if a module M_R has a composition series, then the length and the (simple) factors of the composition series do not depend on the composition series itself, but are determined uniquely by the module M_R. They are called the *length* (or *composition length*) $\ell(M_R)$ and the *composition factors* of the module, respectively. Modules that have a composition series are also called modules of *finite composition length* (or of *finite length*).

For instance, the $\mathbb{Z}$-module $\mathbb{Z}/6\mathbb{Z}$ is a module of finite composition length (it is a module with finitely many elements!) Both

$$0 = 6\mathbb{Z}/6\mathbb{Z} \subseteq 2\mathbb{Z}/6\mathbb{Z} \subseteq \mathbb{Z}/6\mathbb{Z}$$

and

$$0 = 6\mathbb{Z}/6\mathbb{Z} \subseteq 3\mathbb{Z}/6\mathbb{Z} \subseteq \mathbb{Z}/6\mathbb{Z}$$

2.13 Series of Modules: Modules of Finite Composition Length

are composition series, so that $\mathbb{Z}/6\mathbb{Z}$ is a module of composition length 2. The two compositions factors of $\mathbb{Z}/6\mathbb{Z}$ are isomorphic one to $\mathbb{Z}/2\mathbb{Z}$ and the other to $\mathbb{Z}/3\mathbb{Z}$, and they do not appear in the same order in the two composition series written above.

> **Proposition 2.13.5**
> Let R be a ring. A module M_R is of finite composition length if and only if M_R is both artinian and noetherian.

Proof Assume that M_R has finite composition length n. We will argue by induction on n. If $n = 0$, then $M_R = 0$ is both artinian and noetherian. If $n > 0$, let

$$0 = M_0 \subset M_1 \subset \cdots \subset M_n = M_R$$

be a composition series of M_R. Then M_{n-1} has composition length $n-1$, so that M_{n-1} is both artinian and noetherian by the inductive hypothesis. The factor module M_R/M_{n-1} is simple, hence it is both artinian and noetherian. Applying Proposition 2.11.4 to the exact sequence $0 \to M_{n-1} \to M_R \to M_R/M_{n-1} \to 0$, we get that M_R is both artinian and noetherian.

Suppose now that M_R is both artinian and noetherian. We will construct by induction a composition series for M_R. Set $M_0 := 0$. If M_n has been constructed, either $M_n = M_R$ and the chain ends, or we may consider the non-empty set $\{X \leq M_R \mid M_n \subset X\}$ and choose a minimal element M_{n+1}, using the fact that M_R is artinian. Since M_R is noetherian, the strictly ascending chain must end at some $n \geq 0$. The resulting series $0 = M_0 \subset M_1 \subset \cdots \subset M_n = M_R$ is a composition series for M_R, because the factors M_n/M_{n-1} are simple by construction. $\square$

> **Corollary 2.13.6**
> If
>
> $$0 \longrightarrow A_R \xrightarrow{\alpha} B_R \xrightarrow{\beta} C_R \longrightarrow 0$$
>
> is a short exact sequence, B_R is of finite length if and only if both A_R and C_R are of finite length. In this case, $\ell(B_R) = \ell(A_R) + \ell(C_R)$, and the composition factors of B_R are those of A_R plus those of C_R.

Proof The module B_R is of finite length if and only if B_R is both artinian and noetherian (Proposition 2.13.5) if and only if both A_R and C_R are both artinian and noetherian (Proposition 2.11.4) if and only if A_R and C_R are of finite length (Proposition 2.13.5).

Let

$$0 = A_0 \subset A_1 \subset \cdots \subset A_s = A_R$$

and

$$0 = C_0 \subset C_1 \subset \cdots \subset C_t = C_R$$

be composition series of A_R and C_R respectively. We may then consider the series of B_R

$$0 = \alpha(A_0) \subset \alpha(A_1) \subset \cdots \subset \alpha(A_s) = \alpha(A_R) =$$
$$= \ker \beta = \beta^{-1}(C_0) \subset \beta^{-1}(C_1) \subset \cdots \subset \beta^{-1}(C_t) = \beta^{-1}(C_R) = B_R. \qquad (2.12)$$

For all $i = 1, \ldots, s$, $\alpha(A_i)/\alpha(A_{i-1}) \cong A_i/A_{i-1}$ is simple, and for all $j = 1, \ldots, t$, $\beta^{-1}(C_j)/\beta^{-1}(C_{j-1}) \cong C_j/C_{j-1}$ is simple. Thus (2.12) is a composition series of length $s + t$ for B_R and its composition factors are those of A_R together with those of C_R, as we wanted to prove. □

Corollary 2.13.7
If $M_1, \ldots, M_n$ are modules of finite length, then $M = M_1 \oplus \cdots \oplus M_n$ is a module of finite length and $\ell(M) = \ell(M_1) + \cdots + \ell(M_n)$.

Proof For $n = 1$, there is nothing to prove. So assume $n > 1$ and the statement true for $n - 1$. Set $N := M_1 \oplus \cdots \oplus M_{n-1}$, so that $M = N \oplus M_n$. Then N is of finite length by induction, and by considering the canonical split short exact sequence $0 \to N \to M \to M_n \to 0$, we get by the previous corollary that M is of finite length and $\ell(M) = \ell(N) + \ell(M_n)$. By induction, this is equal to $\ell(M_1) + \cdots + \ell(M_n)$, as we wanted to prove. □

Corollary 2.13.8
The following conditions are equivalent for a semisimple module S:

  (i) *S is the direct sum of finitely many simple submodules.*
  (ii) *S is of finite composition length.*
  (iii) *S is artinian.*
  (iv) *S is noetherian.*

2.13 Series of Modules: Modules of Finite Composition Length

Proof

(i) $\Rightarrow$ (ii) Simple modules are modules of finite composition length 1. If S is a direct sum of n simple submodules, then S has composition length n by Corollary 2.13.7.

(ii) $\Rightarrow$ (iii) and (ii) $\Rightarrow$ (iv) follow from Proposition 2.13.5.

(iii) $\Rightarrow$ (i) Assume (i) false. Since S is semisimple, S is the direct sum of a family $\{S_\lambda \mid \lambda \in \Lambda\}$ of simple submodules of S by Theorem 2.10.7, and by our assumption Λ must be infinite. Hence it is possible to fix a countable set of distinct indices λ_i ($i \geq 0$) in the set Λ. Set $M_n := \sum_{i \geq n} S_{\lambda_i}$ for every $n \geq 0$. Then the modules M_n form a properly descending chain of submodules.

(iv) $\Rightarrow$ (i) Argue as for the proof of the previous implications. Now the submodules $M'_n := \sum_{i=1}^n S_{\lambda_i}$ form a properly ascending chain of submodules of S.

$\square$

Right Artinian Rings 3

In the previous Chapter, we have studied modules introducing seven important classes of modules: free modules, projective modules, simple modules, semisimple modules, noetherian modules, artinian modules and modules of finite composition length. In this Chapter, we will consider some classes of rings. More precisely, we will see that there is a hierarchy {division rings} $\subseteq$ {semisimple artinian rings} $\subseteq$ {right artinian rings} $\subseteq$ {right noetherian rings}. We begin by presenting the class of semisimple artinian rings.

3.1 Semisimple Artinian Rings

In the next lemma, we deal with minimal right ideals of a ring R. "Minimal right ideal" means minimal element in the set of all non-zero right ideals of R partially ordered by set inclusion. Clearly a right ideal I of a ring R is a minimal right ideal of R if and only if it is a simple submodule of the right module R_R.

> **Lemma 3.1.1**
> Let I be a minimal right ideal of a ring R. Then either $I^2 = 0$ or $I = eR$ for some idempotent $e \in R$.

Proof Let I be a minimal right ideal of R, and suppose $I^2 \neq 0$. Then there exists an element $a \in I$ with $aI \neq 0$. Also, aI is a submodule of I_R, so that $aI = I$ because I_R is simple. Therefore $a = ae$ for a suitable $e \in I$. Consider $J := \{i \in I \mid ai = 0\}$. This is clearly a right ideal of R contained in I. Now $a = ae \neq 0$, so that $e \notin J$, hence we have that $J \neq I$. Thus $J = 0$ because I is simple. Moreover $a(e^2 - e) = aee - ae = ae - ae = 0$, so that $e^2 - e \in J = 0$. It follows that $e^2 = e$, that is, e is idempotent. Now $e \neq 0$ because $a \neq 0$, and $eR \leq I$. Therefore $eR = I$, because I is simple. □

A ring R is *semisimple artinian* if it is right artinian and has no non-zero nilpotent right ideal. To be more precise, we should call such a ring a *right semisimple artinian* ring, because it is defined relatively to the structure of the right module R_R and to right ideals. Also, we should define *left semisimple artinian* rings symmetrically. But as a consequence of the Artin-Wedderburn Theorem 3.2.1, it will follow that a ring is right semisimple artinian if and only if it is left semisimple artinian, so that a reference to the side is useless. In order not to have a too heavy terminology, we call the rings just defined semisimple artinian, without any reference to the right side.

Theorem 3.1.2
The following conditions are equivalent for a ring $R \neq 0$.

(i) *Every right R-module is projective.*
(ii) *Every short exact sequence of right R-modules splits.*
(iii) *Every right R-module is semisimple.*
(iv) *The module R_R is semisimple.*
(v) *The ring R is semisimple artinian.*

Proof
(i) ⇒ (ii) follows immediately from Proposition 2.6.3 ((i) ⇒ (ii)).
(ii) ⇒ (iii) follows from the remark before the statement of Lemma 2.10.4.
(iii) ⇒ (iv) is trivial.
(iv) ⇒ (v) Most of the proof of this implication was seen implicitly in Exercise 2.9.7, but we repeat the argument here for completeness. Assume R_R semisimple. Then R_R is a direct sum of simple right modules: $R = \bigoplus_{\lambda \in \Lambda} S_\lambda$, where the S_λ's are simple right modules, that is, minimal right ideals. Then $1_R \in R = \bigoplus_{\lambda \in \Lambda} S_\lambda$, so that 1_R is a finite sum of elements in the S_λ's in a unique way. That is, there exists a finite subset F of Λ and non-zero elements $s_\lambda \in S_\lambda$ for every $\lambda \in F$ such that $1_R = \sum_{\lambda \in F} s_\lambda$. Let us prove that $R = \bigoplus_{\lambda \in F} S_\lambda$.
Clearly $R \supseteq \bigoplus_{\lambda \in F} S_\lambda$. To prove the other inclusion, fix an element $r \in R$. Then

$$r = 1_R \cdot r = \sum_{\lambda \in F} s_\lambda r \in \bigoplus_{\lambda \in F} S_\lambda.$$

Thus R_R is a finite direct sum of simple submodules, so that R_R is of finite composition length, hence artinian, i.e., the ring R is right artinian.
Now fix a nilpotent right ideal I of R. We must show that $I = 0$. Since R_R is semisimple, I is a direct summand of R_R. By Proposition 2.9.3, there exists an

3.1 Semisimple Artinian Rings

idempotent element $e \in R$ with $I = eR$. Then I nilpotent implies e nilpotent. But the only element that is both nilpotent and idempotent is zero.

(v) $\Rightarrow$ (iv) Assume that (v) holds and (iv) does not. That is, R is a semisimple artinian ring, but R_R is not a semisimple module. In particular, the set $\mathcal{F}$ of all right ideals of R that are not semisimple modules is non-empty. Since R is right artinian, the set $\mathcal{F}$ has a minimal element I. In particular, $I \neq 0$ and I is not a simple module. Using the fact that R is right artinian again, we see that there exists a minimal right ideal J with $0 < J < I$. Since 0 is the unique nilpotent right ideal of R, any minimal right ideal of R is a direct summand of R (Lemma 3.1.1). Thus J is a direct summand of I (Exercise 1.8.1), i.e., $I = J \oplus C$ for some non-zero right ideal C of R contained in I. By the minimality of I both J and C are semisimple, so that their direct sum I is semisimple, a contradiction.

(iv) $\Rightarrow$ (i) Assume that (iv) holds, and consider an arbitrary right R-module M_R. There exists a short exact sequence

$$0 \to K \hookrightarrow F_R \to M_R \to 0$$

with F_R free. Since R_R is semisimple, it is a direct sum of simple modules. Hence F_R is a direct sum of simple modules, i.e., F_R is semisimple. Thus the short exact sequence splits. So M is isomorphic to a direct summand of F, hence M is projective.

$\square$

The Ring of $n \times n$ Matrices Over a Division Ring

In this Section, we will describe the structure of the ring of all $n \times n$ matrices with entries in a division ring.

Let D be a division ring, $n \geq 1$ an integer, and let $R := \mathbb{M}_n(D)$ be the ring of all $n \times n$ matrices with entries in D. We have already seen in Exercise 1.1.11 that R is a simple ring. For every $i, j = 1, 2, \ldots, n$, let $E_{i,j}$ be the matrix with the (i, j) entry equal to 1, and 0 in all the other entries. Notice that the $E_{i,i}$'s are idempotent elements of R, $E_{1,1} + \cdots + E_{n,n} = 1$ and that $E_{i,i}E_{j,j} = 0$ for $i \neq j$. In particular, by Proposition 2.9.3, the principal right ideal $E_{i,i}R$ generated by $E_{i,i}$ is a direct summand of R_R. More precisely, it is easily seen that $E_{i,i}R$ is the set of all $n \times n$ matrices with entries in D, that are 0 on all rows except for the i-th row, and with arbitrary entries in D on the i-th row. Also, by Exercise 2.9.4,

$$R_R = E_{1,1}R \oplus E_{2,2}R \oplus \cdots \oplus E_{n,n}R$$

The modules $E_{i,i}R$ are isomorphic. For instance, an isomorphism $E_{1,1}R \to E_{i,i}R$ is given by left multiplication by the matrix $E_{i,1}$. Moreover, the module $E_{1,1}R$ is simple (to see this, show that it is generated by any of its non-zero elements). Thus $R_R = E_{1,1}R \oplus \cdots \oplus E_{n,n}R$ is a direct sum of n simple isomorphic modules, in particular R_R is a semisimple right R-module of composition length n. For instance,

$$0 \subset E_{1,1}R \subset E_{1,1}R \oplus E_{2,2}R \subset \cdots \oplus_{i=1}^{n-1} E_{i,i}R \subset R_R \tag{3.1}$$

is a composition series for the module R_R. Thus *R is a semisimple artinian ring*. It follows, in particular, that *R* is right artinian and right noetherian.

Matrix transposition $t: A \mapsto A^t$ is a ring isomorphism $t: \mathbb{M}_n(D) \to (\mathbb{M}_n(D^{\mathrm{op}}))^{\mathrm{op}}$. Therefore *R* is isomorphic to the opposite ring of $\mathbb{M}_n(D^{\mathrm{op}})$, where D^{op} is also a division ring. Thus all properties we have seen on the right also hold on the left. Also the category R−Mod, which is equivalent to the category Mod−R^{op}, is equivalent to the category Mod−$(\mathbb{M}_n(D^{\mathrm{op}}))$.

We have that the left ideal $RE_{i,i}$ *is the set of all* $n \times n$ *matrices with entries in D, that are 0 on all columns except for the i-th column, and with arbitrary entries in D on the i-th column*. We have

$$_RR = RE_{1,1} \oplus RE_{2,2} \oplus \cdots \oplus RE_{n,n},$$

and *the left ideals* $RE_{1,1}, \ldots, RE_{n,n}$ *are isomorphic simple modules*. In particular $_RR$ is a semisimple left R-module of composition length n. Hence *R is left artinian and left noetherian* as well.

Also, *every simple right R-module is isomorphic to* $E_{1,1}R$. To see this, fix an arbitrary simple right R-module S_R. We know that $S_R \cong R_R/M$ for some maximal right ideal *M* of *R*. The series

$$0 \subset M \subset R_R \tag{3.2}$$

can be refined to a composition series of R_R. In particular, the simple factor S_R of the series (3.2) must be isomorphic to one of the composition factors of R_R, which are *n* simple modules all isomorphic to $E_{1,1}R$. Thus $S_R \cong E_{1,1}R$.

Let us prove now that *the endomorphism ring* $\mathrm{End}(E_{1,1}R)$ *of the simple module* $E_{1,1}R$ *is isomorphic to the division ring D*. By Exercise 2.9.6, we know that $\mathrm{End}(E_{1,1}R) \cong E_{1,1}RE_{1,1}$. It is easily seen that $E_{1,1}RE_{1,1}$ consists of all $n \times n$ matrices whose entries are all zero, except for the (1, 1) entry, which can be any element of *D*. It is clear that this set, with the operations induced by the operations of *R*, is a ring isomorphic to the division ring *D*.

Now if M_R is any right R-module, then M_R is semisimple by Theorem 3.1.2. Hence M_R is a direct sum of simple submodules. But all simple right R-modules are isomorphic to $E_{1,1}R$. Thus we have seen that *every right R-module is isomorphic to a direct sum* $E_{1,1}R^{(X)}$ *for some set X, whose cardinality is uniquely determined* by Proposition 2.10.8.

Notice that the tensor product $E_{1,1}R \otimes_R RE_{1,1}$ is an abelian group that does not have an R-module structure. For instance, assume that the division ring *D* is a finite field with *q* elements and $n = 2$. We have seen in Exercise 2.9.8 that $E_{1,1}R \otimes_R$

3.1 Semisimple Artinian Rings

$RE_{1,1} \cong E_{1,1}RE_{1,1}$, so that $E_{1,1}R \otimes_R RE_{1,1} \cong D$ has q elements in this case. But every right R-module is isomorphic to a direct sum of copies of $E_{1,1}R$, which has q^2 elements. Hence every finite right R-module has q^{2t} elements for some nonnegative integer t. Thus no right R-module can have q elements. This proves that $E_{1,1}R \otimes_R RE_{1,1}$ cannot be endowed with a right R-module structure. Similarly, it cannot be endowed with a left R-module structure. Furthermore, one can show in a similar way that $\operatorname{Hom}(E_{1,1}R, E_{1,1}R)$ cannot be endowed with a right R-module structure or a left R-module structure.

We have just seen that a matrix ring with entries in a division ring is a semisimple artinian ring. This is true for finite direct product of such matrix rings, as we show in the next example.

Example 3.1.3
Let $t, n_1, \ldots, n_t \geq 1$ be integers and $D_1, \ldots, D_t$ division rings. The ring $R = M_{n_1}(D_1) \times \cdots \times M_{n_t}(D_t)$ is a semisimple artinian ring.

Proof For all $\ell = 1, \ldots, t$ and for all $i, j = 1, \ldots, n_\ell$, set

$$e_{i,j}^{(\ell)} := (\underbrace{0, \ldots, 0}_{\ell-1}, E_{i,j}, \underbrace{0, \ldots, 0}_{t-\ell}),$$

where $E_{i,j} \in M_{n_\ell}(D_\ell)$ is defined as in the previous example. If $(A_1, \ldots, A_t) \in R$, we have

$$e_{i,j}^{(\ell)}(A_1, \ldots, A_t) = (\underbrace{0, \ldots, 0}_{\ell-1}, E_{i,j}A_\ell, \underbrace{0, \ldots, 0}_{t-\ell})$$

so that

$$\bigoplus_{i=1}^{n_\ell} e_{i,i}^{(\ell)} R = \left\{ (\underbrace{0, \ldots, 0}_{\ell-1}, A, \underbrace{0, \ldots, 0}_{t-\ell}) \mid A \in M_{n_\ell}(D_\ell) \right\}$$

thus

$$R_R = \bigoplus_{\ell=1}^{t} \bigoplus_{i=1}^{n_\ell} e_{i,i}^{(\ell)} R.$$

Since the modules $e_{i,i}^{(\ell)} R$ are simple (cf. the previous example), R_R is a semisimple module, hence R is a semisimple artinian ring by Theorem 3.1.2. □

Exercise 3.1.4
In this exercise, we will describe the right ideals and the left ideals of the ring of matrices $\mathbb{M}_n(D)$. Here D is a division ring and $n \geq 1$ is an integer. Clearly, $\mathbb{M}_n(D)$ is isomorphic to the ring $R := \operatorname{End}(V_D)$, where V_D is any n-dimensional right vector space over the division ring D.

(1) For every subspace W_D of V_D, set $I_W := \{ f \in R \mid f(V_D) \subseteq W_D \}$. Show that I_W is a right ideal of R.
(2) Show that there is an order-preserving one-to-one correspondence $\mathcal{L}(V_D) \to \mathcal{L}(R_R)$, $W_D \mapsto I_W$, between the lattice $\mathcal{L}(V_D)$ of all vector subspaces of V_D and the lattice $\mathcal{L}(R_R)$ of all right ideals of R.
(3) For every subspace W_D of V_D, set $J_W := \{ f \in R \mid f(W_D) = 0 \}$. Show that J_W is a left ideal of R.
(4) Show that there is an order-reversing one-to-one correspondence $\mathcal{L}(V_D) \to \mathcal{L}(_R R)$, $W_D \mapsto J_W$, between the lattice $\mathcal{L}(V_D)$ of all vector subspaces of V_D and the lattice $\mathcal{L}(_R R)$ of all left ideals of R.

Exercise 3.1.5
Let R be a ring with identity. Prove that there is an isomorphism between the lattice of all submodules of R_R^n and the lattice of all left ideals of the ring of matrices $\mathbb{M}_n(R)$. [*Hint: For every left ideal I of $\mathbb{M}_n(R)$ and every submodule N of R_R^n define*

$$\Phi(I) = \{ (r_1, \ldots, r_n) \in R_R^n = \mathbb{M}_{1 \times n}(R) \mid \begin{pmatrix} r_1 & \ldots & r_n \\ 0 & \ldots & 0 \\ \vdots & & \\ 0 & \ldots & 0 \end{pmatrix} \in I \}$$

and

$$\Psi(N) = \begin{pmatrix} N \\ \vdots \\ N \end{pmatrix} .]$$

Example of Ring That Is Left Hereditary But Not Right Hereditary

Examples of rings that are left hereditary but not right hereditary can be found in Kaplansky [17], Small [27, 28], Cohn [6, 15]. The following exercise is taken from [23, p. 62].

Exercise 3.1.6
Set $R := \begin{pmatrix} \mathbb{Z} & 0 \\ \mathbb{Q} & \mathbb{Q} \end{pmatrix}$. Show that:

(i) $A := \begin{pmatrix} 0 & 0 \\ \mathbb{Q} & 0 \end{pmatrix}$ and $B := \begin{pmatrix} 0 & 0 \\ 0 & \mathbb{Q} \end{pmatrix}$ are minimal left ideals of R.
(ii) $_R B$ is projective.
(iii) $_R B \cong {_R A}$.
(iv) The left ideals of R containing $_R A \oplus {_R B}$ are exactly the sets $C_z := \begin{pmatrix} z\mathbb{Z} & 0 \\ \mathbb{Q} & \mathbb{Q} \end{pmatrix}$ for some $z \in \mathbb{Z}$.
(v) The left ideals C_z are isomorphic to $_R R$ for every non-zero $z \in \mathbb{Z}$.
(vi) Show that R is left hereditary. [Hint: Let I be a left ideal. If $I \not\supseteq A$, then $I \cap A = 0$, so that $I \oplus A$ is a left ideal of R that contains A and it suffices to show that $I \oplus A$ is projective. Thus we can suppose that I contains A. If $I \not\supseteq B$, then...]
(vii) Show that A and $D := A \oplus B$ are two-sided ideals of R.

(viii) Show that $\bigcap_{n \geq 1} nA = A$ and $\bigcap_{n \geq 1} nR = D$.
(ix) Show that if $F_R = \oplus_{\lambda \in \Lambda} x_\lambda R$ is a free right R-module, then

$$\bigcap_{n \geq 1} nF_R = \oplus_{\lambda \in \Lambda} x_\lambda D = F_R D.$$

(x) Show that R is not right hereditary. [Hint: It suffices to show that A_R is not projective. Assume the contrary, so that there exists a free right R-module $F_R = \oplus_{\lambda \in \Lambda} x_\lambda R$ with a direct-sum decomposition $F_R = P_R \oplus Q_R$ with $P_R \cong A_R$. Then $P_R = \bigcap_{n \geq 1} nP_R \subseteq \bigcap_{n \geq 1} nF_R = \oplus_{\lambda \in \Lambda} x_\lambda D = F_R D = P_R D + Q_R D$. But $P_R D \cong A_R D = \begin{pmatrix} 0 & 0 \\ \mathbb{Q} & 0 \end{pmatrix} \begin{pmatrix} 0 & 0 \\ \mathbb{Q} & \mathbb{Q} \end{pmatrix} = 0$. Thus $P_R \subseteq Q_R D \subseteq Q_R$. Hence $P_R \subseteq P_R \cap Q_R = 0$, contradiction.]

3.2 The Artin-Wedderburn Theorem

Theorem 3.2.1 (Artin-Wedderburn)
A ring R is semisimple artinian if and only if there exist integers $t, n_1, \ldots, n_t \geq 1$ and division rings $D_1, \ldots, D_t$ such that

$$R \cong M_{n_1}(D_1) \times \cdots \times M_{n_t}(D_t). \tag{3.3}$$

Moreover, if R is semisimple artinian, the integers $t, n_1, \ldots, n_t$ in the decomposition (3.3) are uniquely determined by R and $D_1, \ldots, D_t$ are determined by R up to ring isomorphism.

Proof If the ring R is a direct product of matrix rings as in Eq. (3.3), then R is a semisimple artinian ring, as we have seen in Example 3.1.3.

We will now assume that R is semisimple artinian and prove that R is isomorphic to such a direct product of matrix rings. By Theorem 3.1.2, R_R is a semisimple module, hence a direct sum of simple modules. Moreover, it is a finite direct sum of simple modules (Exercise 2.9.7). Thus $R_R \cong S_1^{n_1} \oplus \cdots \oplus S_t^{n_t}$ say, with $S_i \not\cong S_j$ if $i \neq j$ (we have grouped together isomorphic simple modules). We remark that if $i \neq j$, then

$$\mathrm{Hom}_R(S_i^{n_i}, S_j^{n_j}) \cong \mathrm{Hom}_R(S_i, S_j)^{n_i n_j} = 0 \tag{3.4}$$

by Proposition 1.8.3 and by the fact that $\mathrm{Hom}_R(S_i, S_j) = 0$ because S_i and S_j are non-isomorphic simple modules. Now, $R \cong \mathrm{End}_R(R_R)$, and we have, by Eq. (1.6),

$$\mathrm{End}_R(R_R) \cong \begin{pmatrix} \mathrm{Hom}_R(S_1^{n_1}, S_1^{n_1}) & \cdots & \mathrm{Hom}_R(S_t^{n_t}, S_1^{n_1}) \\ \mathrm{Hom}_R(S_1^{n_1}, S_2^{n_2}) & \cdots & \mathrm{Hom}_R(S_t^{n_t}, S_2^{n_2}) \\ \vdots & & \vdots \\ \mathrm{Hom}_R(S_1^{n_1}, S_t^{n_t}) & \cdots & \mathrm{Hom}_R(S_t^{n_t}, S_t^{n_t}) \end{pmatrix}.$$

By Eq. (3.4) the non-diagonal entries are zero and, by the isomorphism (1.7) on page 34, the diagonal entries are isomorphic to $\mathbb{M}_{n_\ell}(\mathrm{End}_R(S_\ell))$. Thus, if we set $D_\ell := \mathrm{End}_R(S_\ell)$, which is a division ring by Schur's lemma, we see that the ring $\mathrm{End}_R(R_R)$ is isomorphic to the ring

$$\begin{pmatrix} M_{n_1}(D_1) & 0 & \cdots & 0 \\ 0 & M_{n_2}(D_2) & & 0 \\ \vdots & & \ddots & \\ 0 & 0 & \cdots & M_{n_t}(D_t) \end{pmatrix} \cong M_{n_1}(D_1) \times \cdots \times M_{n_t}(D_t).$$

The integer t is the number of simple right R-modules up to isomorphism. In fact, there are at least t pairwise non-isomorphic simple right R-modules $S_1, \ldots, S_t$. We show that these are the only ones up to isomorphism. In fact, if S_R is simple, then $S_R \cong R/M$ for some maximal right ideal $M \subset R_R$; if we refine $0 \subset M \subset R_R$ to a composition series for R_R we see that R/M is a composition factor of R_R. By the decomposition of R_R into simple modules we can easily construct a composition series with composition factors $S_1, \ldots, S_t$. By Jordan-Hölder, it follows that $S_R \cong S_\ell$ for some $\ell = 1, \ldots, t$.

Moreover, n_ℓ is the number of simple summands of R_R isomorphic to S_ℓ, so that it is uniquely determined by R (cf. Proposition 2.10.8).

Since $D_\ell \cong \mathrm{End}_R(S_\ell)$, it follows that $D_1, \ldots, D_t$ are determined by the simple direct summands of R_R, hence by R_R, up to ring isomorphism. $\square$

More precisely, we have just seen that semisimple artinian rings R are completely determined, up to isomorphism, by a finite non-empty indexed set $\{(D_1, n_1), (D_2, n_2), \ldots, (D_t, n_t)\}$, where the n_i are positive integers and the D_i are division rings determined up to ring isomorphism.

3.2 The Artin-Wedderburn Theorem

Corollary 3.2.2
Let R be a ring. The following conditions are equivalent.

1. *R is semisimple artinian.*
2. *R^{op} is semisimple artinian.*
3. *R is left artinian and does not have any non-zero nilpotent left ideals.*

Proof Assume (1). Then $R \cong M_{n_1}(D_1) \times \cdots \times M_{n_t}(D_t)$ for suitable integers $t, n_1, \ldots, n_t \geq 1$ and suitable division rings $D_1, \ldots, D_t$. It follows that $R^{op} \cong M_{n_1}(D_1)^{op} \times \cdots \times M_{n_t}(D_t)^{op}$. Since the transpose is an isomorphism $M_{n_\ell}(D_\ell) \to M_{n_\ell}(D_\ell^{op})^{op}$ for all $\ell = 1, \ldots, t$, it follows that $R^{op} \cong M_{n_1}(D_1^{op}) \times \cdots \times M_{n_t}(D_t^{op})$, and the rings D_i^{op} are division rings, so that R^{op} is semisimple artinian and (2) holds. Conversely, if R^{op} is semisimple artinian, then $R \cong (R^{op})^{op}$ is semisimple artinian. Hence (1) and (2) are equivalent.

The equivalence of (2) and (3) follows from the fact that the lattices $\mathcal{L}(R^{op}_{R^{op}})$ and $\mathcal{L}(_R R)$ coincide, because an additive subgroup of R is a submodule of $_R R$ if and only if it is a submodule of $R^{op}_{R^{op}}$. □

Exercise 3.2.3
We have already seen direct product of rings in a number of previous exercises and results. We will now pause for a moment to analyze this concept.

If R and S are rings, their *(external) direct product* is the set of all pairs (x, y), with $x \in R$ and $y \in S$, and with the operations defined component-wise:

$$(x, y) + (x', y') = (x + x', y + y'), \qquad (x, y)(x', y') = (xx', yy')$$

for every $x, x' \in R$, $y, y' \in S$. The direct product of R and S is denoted by $R \times S$. More generally, given an arbitrary family $\{ R_\lambda \mid \lambda \in \Lambda \}$ of rings, it is possible to define the direct product of the family, taking the cartesian product with the two operations defined component-wise.

Now let R be a ring. Suppose that R decomposes as a direct sum $R = G \oplus H$ as an additive group. Then, any two elements $r, r' \in R$ can be written in a unique way as $r = g + h$, $r' = g' + h'$, with $g, g' \in G$, $h, h' \in H$, and one has that $r + r' = (g + g') + (h + h')$ and $rr' = gg' + gh' + hg' + hh'$. If we want the multiplication to be also defined component-wise with respect to the direct-sum decomposition $R = G \oplus H$, we must necessarily have that G and H are multiplicatively closed subsets of R and $gh' + hg' = 0$ for every $g, g' \in G$, $h, h' \in H$ (equivalently, $gh = 0$ and $hg = 0$ for every $g \in G$, $h \in H$, because $gh = (g, 0)(0, h) = (0, 0)$ for every $g \in G$, $h \in H$). Under these hypotheses, we have that G and H turn out to be two-sided ideals of R because, for instance, $gr' = g(g' + h') = gg' \in G$.

Therefore, let R be a ring and I, J be two two-sided ideals of R. It is now natural to say that the ring R is the *internal direct product* of its two-sided ideals I and J if R is the direct sum $R = I \oplus J$ of I and J as additive groups. In fact, in this case, the addition is defined component-wise, and so is the multiplication as well, because $IJ, JI \subseteq I \cap J = 0$, so that

$$(i + j)(i' + j') = ii' + ij' + ji' + jj' = ii' + jj',$$

with $ii' \in I$ and $jj' \in J$ for every $i, i' \in I$ and every $j, j' \in J$.

Prove that if R is the internal direct product of two non-zero two-sided ideals I and J, then:
(a) I and J, with the two operations induced by the two operations of R, are two rings with identity.
(b) The mapping $\varphi: I \times J \to R$, defined by $\varphi(i, j) = i + j$ for every $i \in I$, $j \in J$, is a ring isomorphism between the external direct product of rings $I \times J$ and R.

Conversely, prove that:
(c) If $R := R_1 \times R_2$ is the external direct product of two rings R_1, R_2 and $\pi_i: R \to R_i$ is the canonical projection for $i = 1, 2$, then R is the internal direct product of the two non-zero ideals $\ker(\pi_i)$ ($i = 1, 2$) and the restrictions $\pi_1|_{\ker(\pi_2)}: \ker(\pi_2) \to R_1$, $\pi_2|_{\ker(\pi_1)}: \ker(\pi_1) \to R_2$ are two ring isomorphisms between $\ker(\pi_2)$ and R_1 and between $\ker(\pi_1)$ and R_2.

From now on, we will not distinguish between internal direct product of rings and external direct product, and write $R \times S$ in both cases. Direct product of rings also satisfies a *Universal Property*, as follows. Let $\{ R_\lambda \mid \lambda \in \Lambda \}$ be a family of rings, $\prod_{\lambda \in \Lambda} R_\lambda$ be its direct product and $\pi_\mu: \prod_{\lambda \in \Lambda} R_\lambda \to R_\mu$, $\mu \in \Lambda$, be the canonical projection. Then, for every ring S and every family $\{ \varphi_\lambda: S \to R_\lambda \mid \lambda \in \Lambda \}$ of ring homomorphisms, there exists a unique ring homomorphism $\varphi: S \to \prod_{\lambda \in \Lambda} R_\lambda$ making all the diagrams

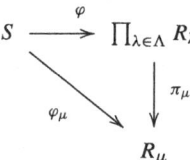

($\mu \in \Lambda$) commute.

Exercise 3.2.4
Show that if R is a ring and $e_1, \ldots, e_n$ are a *complete orthogonal family of central non-zero idempotents* of R, that is, non-zero idempotents of R contained in the center of R such that $e_i e_j = 0$ for $i \neq j$, and $e_1 + \cdots + e_n = 1$, then the ring R is the direct product $R = e_1 R \times \cdots \times e_n R$ (that is, it is the internal direct sum as an abelian group, in which the multiplication is defined component-wise.) Here each $e_i R$ is a ring with identity e_i.

Exercise 3.2.5
Show that the following conditions are equivalent for a ring R and an integer $n \geq 1$:
(a) There exists a complete orthogonal family of n central non-zero idempotents of R.
(b) There exist n non-zero rings $R_1, \ldots, R_n$ such that $R \cong R_1 \times \cdots \times R_n$.

Exercise 3.2.6
(a) Show that the two-sided ideals in a direct product $R = R_1 \times \cdots \times R_n$ of rings are exactly the subsets of R of the form $I_1 \times \cdots \times I_n$ with each I_j a two-sided ideal in R_j.
(b) Deduce from (a) and the Artin-Wedderburn Theorem that every homomorphic image of a semisimple artinian ring is a semisimple artinian ring. [It is also possible to obtain this result in another way, because if R is a semisimple artinian ring and I is a two-sided ideal of R, then R_R is semisimple (Theorem 3.1.2), so that $(R/I)_R$ is semisimple (Lemma 2.10.4). Thus...]

3.3 The Nilradical, Right Artinian Rings, and the Hopkins-Levitzki Theorem

In the next lemma, we have collected some properties concerning nilpotency.

> **Lemma 3.3.1**
> *Let R be a ring.*
>
> (i) *If I is a nilpotent right ideal of R, then RI is a nilpotent two-sided ideal of R.*
> (ii) *If $x \in R$ is nilpotent, then $1 - x$ is invertible.*
> (iii) *Assume $I \subseteq J$, $I \trianglelefteq R$ and $J \leq R_R$. If I is a nil ideal of R and the right ideal J/I of R/I is nil, then J is nil.*
> (iv) *The sum of any family of nil two-sided ideals is a nil two-sided ideal.*

Proof
(i) Notice that RI is the set of all $\sum_{i=1}^{n} r_i x_i$ with $n \geq 1$, $r_i \in R$ and $x_i \in I$ for every $i = 1, \ldots, n$ (see Eq. (2.3) at the beginning of Sect. 2.9.) Thus RI is a two-sided ideal containing I (it is the two-sided ideal generated by I). If I is nilpotent, let $n \geq 1$ be such that $I^n = 0$. Then $(RI)^n = RIRI \ldots RI \subseteq RI \ldots I = RI^n = 0$, because $IR \subseteq I$.

(ii) If $x \in R$ is nilpotent, then $x^n = 0$ for some $n \geq 1$, so that

$$(1-x)(1 + x + x^2 + \cdots + x^{n-1}) = 1 - x^n = 1,$$

and $1 - x$ is invertible.

(iii) Let $x \in J$. Then there exists $n \geq 1$ such that $x^n \in I$, because J/I is nil. Now there exists $m \geq 1$ such that $(x^n)^m = 0$, because I is nil. It follows that $x^{nm} = 0$.

(iv) *First case:* The sum of two nil ideals is a nil ideal.
Let I, J be two nil ideals. Then $(I + J)/I$ is the image of the nil ideal J of R via the ring homomorphism $R \to R/I$. It follows that $(I + J)/I$ is a nil ideal of R/I. Now conclude by (iii).
Second case: The sum of finitely many nil ideals is a nil ideal.
This follows from the previous case by induction on the number of nil ideals.
Third case: The sum of an arbitrary family of nil ideals is a nil ideal.
If Λ is an arbitrary set and I_λ is a nil ideal of R for every $\lambda \in \Lambda$, then $\sum_{\lambda \in \Lambda} I_\lambda$ is nil, because every element $x \in \sum_{\lambda \in \Lambda} I_\lambda$ is a finite sum $x = i_{\lambda_1} + \cdots + i_{\lambda_n}$ for suitable $i_{\lambda_k} \in I_{\lambda_k}$ with $\lambda_k \in \Lambda$ for every $k = 1, \ldots, n$. Thus $x \in I_{\lambda_1} + \cdots + I_{\lambda_n}$, which is a nil ideal by the second case. □

The *nilradical* Nil(R) of a ring R is the sum of all nil two-sided ideals of R. It is the largest nil two-sided ideal of R.

Lemma 3.3.2
For any ring R, Nil(R/Nil(R)) = 0.

Proof By the Correspondence Theorem 1.9.3,

$$\text{Nil}(R/\text{Nil}(R)) = J/\text{Nil}(R)$$

for some ideal J of R containing Nil(R). The ideal $J/\text{Nil}(R)$ is nil. Since Nil(R) is nil, also J is nil, so that $J \subseteq \text{Nil}(R)$, i.e., $J = \text{Nil}(R)$. □

Exercise 3.3.3
(a) Let R be a commutative ring. Show that the nilradical of R is exactly the set of all nilpotent elements of R.
(b) Show that (a) does not necessarily hold for non-commutative rings. As an example, consider the ring $R := M_2(D)$, D a division ring.

The following result is a corollary of Theorem 3.2.1.

Corollary 3.3.4 (The Wedderburn Theorem)
A ring R is simple and right artinian if and only if $R \cong \mathbb{M}_n(D)$ for some integer $n \geq 1$ and some division ring D.

Proof Assume R simple and right artinian. For any right ideal I with $I^n = 0$, we have that RI is a two-sided ideal of R. Clearly, $(RI)^n = 0$, so that RI cannot be the improper ideal R. As R is simple, it follows that $RI = 0$. Thus $I = 0$. This proves that R does not have non-zero nilpotent right ideals. Hence R is semisimple Artinian, so that $R \cong \prod_{i=1}^t \mathbb{M}_{n_i}(D_i)$ for suitable positive integers $t, n_1, \ldots, n_t$ and suitable division rings D_i. If $t \geq 2$, then $\prod_{i=2}^t \mathbb{M}_{n_i}(D_i)$ would be a proper non-zero ideal of $\prod_{i=1}^t \mathbb{M}_{n_i}(D_i)$, so that R would have a proper non-zero ideal, which contradicts the fact that R is simple. Hence $t = 1$ and $R \cong \mathbb{M}_n(D)$.

For the converse, we already know that the matrix ring $\mathbb{M}_n(D)$, with D a division ring, is semisimple artinian, hence right artinian, and simple. □

3.3 The Nilradical, Right Artinian Rings, and the Hopkins-Levitzki Theorem

Theorem 3.3.5
Let R be a right artinian ring and let $N := \mathrm{Nil}(R)$. Then R/N is a semisimple artinian ring and N is the largest nilpotent ideal of R.

Proof Recall that a ring S is semisimple artinian if it is a right artinian ring without non-zero nilpotent right ideals. By Corollary 2.12.3, R right artinian implies R/N right artinian. Now let I/N be a nilpotent right ideal of R/N. Then RI/N is a nilpotent two-sided ideal of R/N (Lemma 3.3.1(i)). Now RI/N and N are nil, so that RI is a nil two-sided ideal of R. Thus $I \subseteq RI \subseteq N$, that is, $I/N = 0$. This proves that R/N is semisimple artinian.

Now we want to show that N is the largest nilpotent ideal of R. Since every nilpotent ideal is nil, it suffices to show that N is nilpotent.

Assume then that N is not nilpotent. Consider the descending chain

$$N \supseteq N^2 \supseteq N^3 \supseteq \ldots$$

As R is right artinian, there exists $k \geq 1$ such that $N^k = N^{k+1}$. Since N is not nilpotent, $N^k \neq 0$. Let

$$I := \mathrm{l.ann}_R(N^k) := \{r \in R \mid rN^k = 0\}$$

be the left annihilator of N^k in R. As N is a two-sided ideal, we also have $I \triangleleft R$. Also, I is a proper ideal of R, because $N^k \neq 0$ (so $1 \notin I$). Therefore

$$\{I' \leq R_R \mid I' \supset I\} \neq \emptyset$$

Since R is right artinian, this set has a minimal element, call it J. Take $a \in J \setminus I$. Then $J = I + aR$ by minimality. Moreover, $I + JN = I + (I + aR)N = I + IN + aRN = I + aN$ because $IN \subseteq I$. Now we have two cases:

Case 1: $JN \not\subseteq I$. In this case, $J \supseteq JN + I \supset I$, so $J = JN + I$ by minimality. Then $a \in JN + I = I + aN$, and $a = i + an$ for suitable $i \in I, n \in N$. Since n is nilpotent, $1 - n$ is invertible, so that $a = i(1-n)^{-1} \in I$, contradiction.

Case 2: $JN \subseteq I$. Then $JN^{k+1} = JNN^k \subseteq IN^k = 0$, so that $JN^{k+1} = 0$. Since $N^k = N^{k+1}$, $JN^k = 0$, and $J \subseteq I = \mathrm{l.ann}_R(N^k)$. Contradiction, because $J \supset I$ properly. $\square$

Theorem 3.3.6 (Hopkins-Levitzki)
A ring R with nilradical N is a right artinian ring if and only if R is right noetherian, N is nilpotent, and R/N is semisimple artinian.

Proof Let R be a right artinian ring, so that R/N is semisimple artinian and N is nilpotent by Theorem 3.3.5. Thus $N^k = 0$ for a suitable $k \geq 0$. Consider the descending chain of right R-modules

$$R = N^0 \supseteq N = N^1 \supseteq N^2 \supseteq \cdots \supseteq N^k = 0.$$

Since R is right artinian, every factor N^i/N^{i+1} is artinian as a right R-module. Moreover $(N^i/N^{i+1})N = N^{i+1}/N^{i+1} = 0$, so that N^i/N^{i+1} is an artinian right R/N-module. Since R/N is semisimple artinian, the artinian R/N-module N^i/N^{i+1} is a semisimple R/N-module, that is, a direct sum of finitely many simple R/N-modules. Thus the R/N-module N^i/N^{i+1} is noetherian. It follows that R_R is noetherian, as desired.

Conversely, assume R right noetherian, N nilpotent and R/N semisimple artinian. Then $N^k = 0$ for some $k \geq 0$ because N is nilpotent. In the descending chain $R = N^0 \supseteq N = N^1 \supseteq N^2 \supseteq \cdots \supseteq N^k = 0$, all factors N^i/N^{i+1} are noetherian right R-modules because R_R is noetherian. They are annihilated by N, hence they are right R/N-modules. Thus the right R/N-modules N^i/N^{i+1} are noetherian. But R/N is semisimple artinian, so that the semisimple noetherian right R/N-modules N^i/N^{i+1} are modules of finite length. Thus the right R-modules N^i/N^{i+1} have finite length, so that R_R has finite length. In particular, R is right artinian. $\square$

Corollary 3.3.7
Every finitely generated right R-module over a right artinian ring R has finite composition length.

Proof Let R be right artinian, so that R is right noetherian by the Hopkins-Levitzki Theorem. Any finitely generated right module M_R is both noetherian and artinian by Corollary 2.11.6, hence it has finite composition length by Proposition 2.13.5. $\square$

3.4 The Radical of a Module

A submodule N of a module M_R is *superfluous* (or *small*, or *inessential*) in M_R if, for every submodule L of M_R, $N + L = M_R$ implies $L = M_R$. To denote that N is superfluous in M_R, we will write $N \leq_s M_R$.

Examples 3.4.1
(1) The only superflous submodule of $\mathbb{Z}_\mathbb{Z}$ is 0. To see this, let I be a non-zero submodule of $\mathbb{Z}_\mathbb{Z}$. Let n be a non-zero element in I and p a prime that does not divide n. Then $I + p\mathbb{Z} = \mathbb{Z}$ and $p\mathbb{Z}$ is proper. Thus I is not superfluous in $\mathbb{Z}$.
(2) In $\mathbb{Z}(p^\infty)$ all proper submodules are superfluous, because the sum of any two proper submodules is a proper submodule.

Exercise 3.4.2
Show that:
 (i) If $K \leq N \leq M_R$, then $N \leq_s M$ if and only if $K \leq_s M$ and $N/K \leq_s M/K$.
 (ii) $K \leq_s M_R$ and $M_R \leq N_R$ imply $K \leq_s N_R$.
 (iii) If $N, N' \leq M_R$, then $N + N' \leq_s M$ if and only if $N \leq_s M$ and $N' \leq_s M$.
 (iv) The zero submodule is always a superfluous submodule of any module M_R, also when $M_R = 0$.
 (v) If $f : M \to M'$ is an R-module morphism and $N \leq_s M$, then $f(N) \leq_s M'$.
 (vi) Assume $K_1 \leq M_1 \leq M$, $K_2 \leq M_2 \leq M$ and $M = M_1 \oplus M_2$. Show that $K_1 \oplus K_2 \leq_s M_1 \oplus M_2$ if and only if $K_1 \leq_s M_1$ and $K_2 \leq_s M_2$.

We will say that an epimorphism $g \colon M_R \to N_R$ is *superfluous* if $\ker g$ is a superfluous submodule of M_R.

Exercise 3.4.3
Show that an epimorphism $g \colon M \to N$ is superfluous if and only if for every module L and every homomorphism $h \colon L \to M$, if gh is onto, then h is onto.

The *radical* $\mathrm{rad}(M_R)$ of a module M_R is the intersection of all maximal submodules of M_R. Note the duality with the definition of socle, which is the sum of all simple (=minimal) submodules of M_R.

Thus if a module M_R does not have maximal submodules, then

$$\mathrm{rad}(M_R) = M_R,$$

because it is the intersection of an empty family of submodules of M_R. [The reader who does not like this kind of conventions can define the radical of a module in the following equivalent way. Since the maximal submodules of M_R are exactly the kernels of the non-zero homomorphisms of M_R into simple modules, we can define $\mathrm{rad}(M_R)$ as the set of all $x \in M$ such that $f(x) = 0$ for every simple right R-module S_R and every $f \in \mathrm{Hom}(M, S)$. Equivalently, $\mathrm{rad}(M)$ is the intersection of all the kernels $\ker(f)$ for every simple right R-module S_R and every $f \in \mathrm{Hom}(M, S)$. If the module M_R has no maximal submodules, then $\mathrm{rad}(M) = M$, because for every simple right R-module S_R, the only homomorphism $M_R \to S_R$ is the zero morphism.]

Example 3.4.4
$\mathrm{rad}(\mathbb{Z}) = 0$, and $\mathrm{rad}(\mathbb{Z}(p^\infty)) = \mathbb{Z}(p^\infty)$.

Lemma 3.4.5
For any module M_R, the submodule $\mathrm{rad}(M_R)$ is the sum of all superfluous submodules of M_R.

Proof In order to prove that $\mathrm{rad}(M_R)$ contains the sum of all superfluous submodules, fix a superfluous submodule N of M_R and a maximal submodule M' of M_R. We must prove that $N \subseteq M'$. Assume the contrary. Then $N + M' = M_R$ because M' is maximal, so that $M' = M_R$ because N is superfluous. Contradiction.

For the inverse inclusion, it suffices to show that if $x \in \mathrm{rad}(M_R)$, then xR is superfluous in M_R. Equivalently, we will prove that if $x \in M_R$ and xR is not superfluous in M_R, then $x \notin \mathrm{rad}(M_R)$. Let x be an element of M_R with xR non-superfluous in M_R, so that there exists a proper submodule L of M_R with $xR + L = M_R$. Then $x \notin L$. Set $\mathcal{F} := \{S \mid S \leq M_R, S \supseteq L, x \notin S\}$. The family $\mathcal{F}$ is non-empty because it contains L. Zorn's Lemma shows that there is a maximal element $\overline{S}$ in $\mathcal{F}$. Then $\overline{S}$ is a maximal submodule of M_R, because if S' is a submodule of M_R that contains $\overline{S}$ properly, then $S' \notin \mathcal{F}$, so that $x \in S'$, whence $S' \supseteq xR + L = M_R$. As x does not belong to the maximal submodule $\overline{S}$ of M_R, it follows that $x \notin \mathrm{rad}(M_R)$, as desired. □

Proposition 3.4.6
For every right module M_R over a ring R,

$$\mathrm{rad}(M_R/\mathrm{rad}(M_R)) = 0.$$

Proof We have that $\mathrm{rad}(M/\mathrm{rad}(M))$ is the intersection of all maximal submodules $\overline{N}$ of $M/\mathrm{rad}(M)$. By the Correspondence Theorem 1.9.3, any such maximal submodule $\overline{N}$ is equal to $N/\mathrm{rad}(M)$ for some maximal submodule N of M containing $\mathrm{rad}(M)$. But every maximal submodule of M contains $\mathrm{rad}(M)$, so that $\mathrm{rad}(M/\mathrm{rad}(M))$ is the intersection of all $N/\mathrm{rad}(M)$ where N ranges in the set of all maximal submodules of M. This is equal to the intersection of all maximal submodules of M, modulo $\mathrm{rad}(M)$, that is, to $\mathrm{rad}(M)/\mathrm{rad}(M) = 0$. □

From Exercise 3.4.2(v) and Lemma 3.4.5 we immediately get that:

Corollary 3.4.7
If $f: M_R \to M'_R$ is a homomorphism of R-modules, then

$$f(\mathrm{rad}(M_R)) \le \mathrm{rad}(M'_R).$$

In particular, $\mathrm{rad}(M_R)$ is a subbimodule of the bimodule $_{\mathrm{End}(M_R)}M_R$.

3.5 The Jacobson Radical of a Ring

The radical of the right R-module R_R is called the *Jacobson radical* of the ring R. It is denoted $J(R)$. Thus $J(R) := \mathrm{rad}(R_R)$ is the intersection of all maximal right ideals of R. Clearly, $J(R)$ is a right ideal of R, because it is defined as an intersection of right ideals.

To be more precise, we should call $\mathrm{rad}(R_R)$ the *right Jacobson radical* of R, but we will see, as a corollary to Proposition 3.5.3(iii), that $\mathrm{rad}(R_R) = \mathrm{rad}(_R R)$ for any ring R.

For every right R-module M_R, the *right annihilator* $\mathrm{r.ann}_R(M_R)$ of M_R is the set of all $r \in R$ such that $Mr = 0$. The right annihilator of any right R-module is a two-sided ideal of R. If $x \in M_R$, the *right annihilator* $\mathrm{r.ann}_R(x)$ of x is the set of all $r \in R$ such that $xr = 0$. The right annihilator of an element x of M_R is a right ideal of R.

Lemma 3.5.1
The Jacobson radical $J(R)$ of any ring R is the intersection of the right annihilators $\mathrm{r.ann}_R(S_R)$ of all simple right R-modules S_R.

Proof If $r \in J(R)$ and x is an element of a simple right R-module S_R, we can consider the right R-module morphism $\varphi: R_R \to S_R$ defined by $\varphi(t) = xt$ for every $t \in R_R$. From $r \in J(R)$, we get that r must be in the kernel of all morphisms $R_R \to S_R$, so that $\varphi(r) = 0$, that is, $xr = 0$. Thus r is in the right annihilator of S_R for every simple module S_R.

Conversely, let r be an element of the ring R with the property that $S_R r = 0$ for every simple right R-module S_R. In order to prove that $r \in J(R)$, fix an arbitrary maximal right ideal M. Then R/M is simple, so that $(R/M)r = 0$. Thus $(1+M)r = 0_{R/M}$, from which $r \in M$. □

Thus $J(R)$ is the set of all the elements $r \in R$ such that $Sr = 0$ for every simple right R-module S. By Lemma 3.5.1, the Jacobson radical of a ring is a two-sided ideal, because it is an intersection of two-sided ideals.

The next result is due to Goro Azumaya.

Proposition 3.5.2 (Nakayama's Lemma)
Let M_R be a finitely generated right module and let N be a submodule of M_R. Then $N + MJ(R) = M$ implies $N = M$.

Notice that Nakayama's Lemma can also be stated as "If M_R is a finitely generated right module, then $MJ(R)$ is superfluous in M_R."

Proof Assume the contrary, i.e., that there exists a finitely generated right module M_R with a proper submodule N such that $N + MJ(R) = M$. Since M_R is finitely generated, the submodule N of M_R is contained in a maximal submodule M' of M_R. Then $M' + MJ(R) = M$. Since M_R/M' is a simple R-module, it is annihilated by $J(R)$, that is, $0 = (M_R/M')J(R) = M_R J(R) + M'/M'$. Thus $M_R J(R) + M' = M'$, a contradiction. □

Proposition 3.5.3
The Jacobson radical $J(R)$ of a ring R can also be described as:

  (i) *The largest superfluous right ideal of R.*
 (ii) *The set of all $x \in R$ such that $1 - xr$ is right invertible for every $r \in R$.*
(iii) *The set of all $x \in R$ such that $1 - rxs$ is invertible for every $r, s \in R$.*

Proof
  (i) Every superfluous right ideal of R is contained in $J(R)$ by Lemma 3.4.5. Moreover, the right ideal $J(R)$ is superfluous in R_R by Nakayama's Lemma.
 (ii) Let x be an element of $J(R)$ and $r \in R$. By (i), $J(R) \leq_s R_R$, so that $xrR \leq_s R_R$. As $xrR + (1 - xr)R = R_R$, it follows that $(1 - xr)R = R_R$, so that $1 - xr$ is right invertible.
      Conversely, let x be an element of R not in $J(R)$. Then x is not contained in a maximal right ideal M of R. Hence $xR + M = R_R$, so that $xr + y = 1$ for some $r \in R$, $y \in M$. Then $1 - xr$ belongs to the proper right ideal M of R, hence $1 - xr$ is not right invertible.
(iii) Clearly, the set described in (iii) is contained in the set described in (ii), that is, in $J(R)$.

3.5 The Jacobson Radical of a Ring

Conversely, fix $x \in J(R)$ and $r, s \in R$, so that $rxs \in J(R)$ because $J(R)$ is a two-sided ideal. By (ii), the element $1 - rxs$ is right invertible, i.e., (∗) $(1 - rxs)y = 1$ for some $y \in R$. Similarly, $-rxsy \in J(R)$ and (ii) imply that (∗∗) $(1 + rxsy)z = 1$ for some $z \in R$. Adding (∗) multiplied by z to (∗∗), one gets that $yz = 1$. Now (∗) and $yz = 1$ show that y is invertible with inverse $1 - rxs = z$. Thus $1 - rxs$ is invertible as well. □

Notice that Condition (iii) is right/left symmetric, so that $J(R)$ can also be described as the intersection of all maximal left ideals of R (i.e., $\operatorname{rad}(R_R) = \operatorname{rad}(_R R)$), or the unique largest superfluous left ideal of R, or the set of all $x \in R$ such that $1 - rx$ is left invertible for every $r \in R$.

As a corollary, we have that every nil ideal is contained in $J(R)$, because for every nilpotent element x, $1 - x$ is invertible. Thus:

Proposition 3.5.4
For every ring R, $\operatorname{Nil}(R) \subseteq J(R)$.

Proposition 3.5.5
Let R be a ring and I be a two-sided ideal of R contained in $J(R)$. Then $J(R/I) = J(R)/I$.

Proof Since $J(R)$ contains I, every maximal right ideal of R contains I. Hence the Jacobson radical $J(R/I)$, which is the intersection of all maximal right ideals of R/I, is the intersection of all M/I, where M ranges in the set of all maximal right ideals of R. This is equal to the intersection of all maximal right ideals M of R, modulo I, that is, $J(R)/I$. □

Corollary 3.5.6
For every ring R, one has that $J(R/J(R)) = 0$.

Proposition 3.5.7
Let R be a semisimple artinian ring. Then $J(R) = 0$.

Proof A semisimple artinian ring R is a direct sum of finitely many simple right R-modules $S_1, S_2, \ldots, S_n$. The right ideals $S_1 \oplus \cdots \oplus S_{i-1} \oplus S_{i+1} \oplus \cdots \oplus S_n$ are

n maximal right ideals of R, and the intersection of these n maximal right ideals is zero. Thus $J(R) = 0$. □

Proposition 3.5.8
Let R be a right artinian ring. Then $J(R) = \text{Nil}(R)$.

Proof It suffices to show that $\text{Nil}(R) \supseteq J(R)$. Since R is right artinian, $R/\text{Nil}(R)$ is semisimple artinian (Theorem 3.3.5), so $J(R/\text{Nil}(R)) = 0$ by the previous proposition. But $\text{Nil}(R) \subseteq J(R)$ (Proposition 3.5.4), so that $0 = J(R/\text{Nil}(R)) = J(R)/\text{Nil}(R)$ by Proposition 3.5.5. Thus $J(R) = \text{Nil}(R)$. □

Exercise 3.5.9
A ring R is a *prime ring* if, for every $a, b \in R$, $aRb = 0$ implies $a = 0$ or $b = 0$.
Show that the following conditions are equivalent for a two-sided ideal I:
(a) The ring R/I is a prime ring.
(b) For every $a, b \in R$, $aRb \subseteq I$ implies $a \in I$ or $b \in I$.
(c) For every pair A, B of two-sided ideals of R, $AB \subseteq I$ implies that either $A \subseteq I$ or $B \subseteq I$.
A proper two-sided ideal I of R is a *prime ideal* if it satisfies one of these equivalent conditions.

Exercise 3.5.10
(a) Show that every integral domain is a prime ring.
(b) Show that if R is a commutative ring, then R is an integral domain if and only if it is a prime ring.
(c) Let D be a division ring, V_D a vector space of dimension at least 2 and $R := \text{End}(V_D)$. Show that R is a prime ring and is not an integral domain.

Exercise 3.5.11
The *prime radical* of a ring R is the intersection $P(R)$ of all prime two-sided ideals of R.
An element x of a ring R is *strongly nilpotent* if, for every infinite sequence x_k, $k \geq 0$, of elements of R such that $x_0 = x$ and $x_{k+1} \in x_k R x_k$ for every $k \geq 0$, there exists an integer n such that $x_n = 0$.
(a) Show that every strongly nilpotent element of a ring is nilpotent.
(b) Show that, for every ring R, $P(R)$ is the set of all strongly nilpotent elements of R. This is a result due to Levitzki.
[Hint: For (a), consider the sequence $x_k = x^{2^k}$.
For (b). If x is an element of R not in $P(R)$, there exists a prime ideal P such that x is not in P. Show that x is not strongly nilpotent considering the sequence x_k defined as follows. Set $x_0 := x$. If $x_k \notin P$ has been defined, then there exists $x_{k+1} \in x_k R x_k \setminus P$, and conclude that x is not strongly nilpotent.
For the converse, assume that $x \in R$ is not strongly nilpotent. Consider a sequence x_k that shows that x is not strongly nilpotent. Let S be the set of these x_k's. Let P be an ideal maximal in the set of all two-sided ideals of R disjoint from S. In order to show that P is a prime ideal that does not contain x, let I and J be ideals not contained in P. Then there exist elements $x_n \in I + P$ and $x_m \in J + P$. If $\ell = \max\{n, m\}$, then x_ℓ belongs to both $I + P$ and $J + P$. Thus $x_{\ell+1} \in x_\ell R x_\ell \subseteq (I + P)(J + P) \subseteq IJ + P$. Now conclude.]

Exercise 3.5.12
Show that the following conditions are equivalent for a ring R:
(a) $P(R) = 0$.
(b) R has no non-zero nilpotent right ideals.
(c) R has no non-zero nilpotent two-sided ideals.

[Hint: (a) $\Rightarrow$ (b), because every nilpotent right ideal is contained in every prime ideal. For (c) $\Rightarrow$ (a), it suffices to show that if (c) holds, then R has no strongly nilpotent non-zero elements. To this end, if x is a non-zero element of R, define a sequence of non-zero elements $x_k \in R$ setting (1) $x_0 := x$; (2) if $x_k \neq 0$ has been defined, then $Rx_k R$ is a non-zero ideal, hence it is not nilpotent, so $Rx_k Rx_k R \neq 0$, hence $x_k Rx_k \neq 0$. Let x_{k+1} be any non-zero element of $x_k R x_k$.]

Exercise 3.5.13
Let R be a ring. Show that the following subsets of R coincide:
(a) The union of all nilpotent right ideals of R.
(b) The sum of all nilpotent right ideals of R.
(c) The sum of all nilpotent two-sided ideals of R.

In particular, this subset of R is a two-sided ideal, called the nilpotent radical of R and denoted by $N(R)$.

Exercise 3.5.14
Show that $N(R) \subseteq P(R) \subseteq \text{Nil}(R) \subseteq J(R)$ for any ring R.

Exercise 3.5.15
Show that $N(R) = P(R) = \text{Nil}(R)$ for any commutative ring R.

Exercise 3.5.16
Show that $N(R) = P(R) = \text{Nil}(R) = J(R)$ for any right artinian ring R.
[Hint: $\text{Nil}(R) = J(R)$ by Proposition 3.5.8. $N(R) = \text{Nil}(R)$ follows from Theorem 3.3.5.]

3.6 Group Representations

Let k be a commutative field and $n \geq 1$ be an integer (the classical case is when k is the field of complex number). Let $\text{GL}_n(k)$ be the *general linear group*, that is, the group of invertible elements of $\mathbb{M}_n(k)$ (i.e., the set of matrices with entries in k having non-zero determinant). It is a group under matrix multiplication. The group $\text{GL}_n(k)$ is isomorphic to the automorphism group of an n-dimensional k-vector space.

Let G be a finite group, not necessarily abelian. As usual, the operation of G will be denoted multiplicatively. A *representation* of G of *degree* n over the field k is a group homomorphism $\varrho \colon G \to \text{GL}_n(k)$. Clearly, given a representation of G of degree n over a field k, G acts as a group of transformations on k^n.

Example 3.6.1
By Cayley's theorem, every finite group G of order n is isomorphic to a subgroup of S_n. If we look at S_n as the group of all bijections of the set G into itself, the embedding $\lambda \colon G \to S_n$ is defined by $\lambda(g)(h) = gh$ for every $g, h \in G$. Every bijection of the set G into itself can be extended by k-linearity to an automorphism of the k-vector space of dimension n having G as a basis. This embeds S_n into $\mathrm{GL}_n(k)$ (As an exercise, the reader could prove that the element corresponding in $\mathrm{GL}_n(k)$ to a permutation $\sigma \in S_n$ is the matrix $\sum_{i=1}^{n} E_{\sigma(i),i} \in \mathrm{GL}_n(k)$). The composition of $\lambda \colon G \to S_n$ and the embedding of S_n into $\mathrm{GL}_n(k)$ is a representation of G of degree n. More precisely, it is an injective homomorphism.

If ϱ_1, ϱ_2 are two representations of G of degree n_1, n_2 respectively, their *direct sum* is the representation $\varrho_1 \oplus \varrho_2$ of G of degree $n_1 + n_2$ defined, for every $g \in G$, by $(\varrho_1 \oplus \varrho_2)(g) = \begin{pmatrix} \varrho_1(g) & 0 \\ 0 & \varrho_2(g) \end{pmatrix}$.

Two representations ϱ and ϱ' of G are said to be *equivalent* if they have the same degree n and there exists $U \in \mathrm{GL}_n(k)$ such that $\varrho'(g) = U^{-1}\varrho(g)U$ for every $g \in G$ (i.e., the images $\varrho'(g)$ and $\varrho(g)$ of g are all conjugate by the same invertible matrix U).

A representation ϱ of G of degree n is *decomposable* if it is equivalent to a representation of the form $\varrho_1 \oplus \varrho_2$ for suitable representations ϱ_1, ϱ_2 of G of degree $< n$. Otherwise, ϱ is said to be *indecomposable*.

A representation ϱ of G of degree n is *reducible* if there exist an integer m with $1 \leq m \leq n-1$ and a representation ϱ' of G equivalent to ϱ such that every matrix $\varrho'(g)$ is a block matrix of the form $\begin{pmatrix} A_g & B_g \\ 0 & C_g \end{pmatrix}$ with A_g an $m \times m$ matrix, B_g an $m \times (n-m)$ matrix, 0 the zero $(n-m) \times m$ matrix, and C_g an $(n-m) \times (n-m)$ matrix. Note that in this case $g \mapsto A_g$ and $g \mapsto C_g$ are representations of G of degree m and $n-m$ respectively.

A representation that is not reducible, is said to be *irreducible*. Clearly, every decomposable representation is reducible.

Now we will state the previous concepts in a more modern terminology. For any group G (possibly infinite) and any field k, we can define the *group algebra* $k[G]$ as follows. It is a k-algebra that, as a vector space over k, has G as a basis. Thus $k[G] := \bigoplus_{g \in G} kg$ is the set of all formal linear combinations $\sum_{g \in G} \lambda_g g$ with the coefficients λ_g elements of k almost all zero. The multiplication $k[G] \times k[G] \to k[G]$ on $k[G]$ is defined by extending by k-bilinearity the multiplication $G \times G \to G$ on the group G, i.e.,

$$(\sum_{g \in G} \lambda_g g)(\sum_{h \in G} \mu_h h) = \sum_{g \in G} \sum_{h \in G} \lambda_g \mu_h (gh). \tag{3.5}$$

3.6 Group Representations

Here we mean that

$$\left(\sum_{g \in G} \lambda_g g\right)\left(\sum_{h \in G} \mu_h h\right) = \sum_{\ell \in G} \nu_\ell \ell,$$

where $\nu_\ell = \sum \lambda_g \mu_h$ and where this sum is indexed in the set of all pairs $(g, h) \in G \times G$ with $gh = \ell$. In this way, $k[G]$ becomes a k-algebra of dimension equal to the order of G. The identity of $k[G]$ is $1_{k[G]} = 1_k \cdot 1_G$.

▶ **Remark 3.6.2** As we know, the k-algebra structure on $k[G]$ can also be defined via a ring morphism of k into the center of $k[G]$. In this case, we have the injective ring homomorphism $k \hookrightarrow k[G]$ that maps every $\lambda \in k$ to $\lambda \cdot 1_G$. Via this ring embedding, we will view k as a subring of $k[G]$. There is also a *monoid* embedding $G \hookrightarrow k[G]$ that maps every $g \in G$ to $1_k \cdot g$. Since k is a subring of $k[G]$, every $k[G]$-module is a k-vector space by restriction of scalars (Exercise 2.3.5), hence has a dimension over k.

Given any representation ϱ of G over k, it is possible to associate to ϱ a left $k[G]$-module M, finite dimensional as a k-vector space. Conversely, given any left module $_{k[G]}M$, finite-dimensional over k, we can construct a representation of G associated to $_{k[G]}M$. This allows us to set up a dictionary between representations of G of a fixed degree n and modules over $k[G]$ which are n-dimensional vector spaces over k. In this way, it will be possible to get results about representations of G from the theory of modules. We will get, for any representation of G over $\mathbb{C}$, a direct-sum decomposition into irreducible representations.

Let $\varrho \colon G \to \mathrm{GL}_n(k)$ be a representation of G of degree n. Consider the n-dimensional vector space $_k k^n$ over k. We will write elements of $_k k^n$ as columns, that is, as $n \times 1$ matrices. As $_k k^n$ is a k-vector space, we have a left scalar multiplication $k \times k^n \to k^n$, which we now extend to a mapping $k[G] \times k^n \to k^n$ setting

$$\left(\sum_{g \in G} \lambda_g g, \begin{pmatrix} \alpha_1 \\ \vdots \\ \alpha_n \end{pmatrix}\right) \mapsto \sum_{g \in G} \lambda_g \varrho(g) \begin{pmatrix} \alpha_1 \\ \vdots \\ \alpha_n \end{pmatrix}.$$

In this way, k^n gets a $k[G]$-module structure.

Conversely, suppose we have a left module $_{k[G]}M$, finite-dimensional over k, of dimension n, say. As $_{k[G]}M_k$ is a bimodule, there is a ring homomorphism $\Lambda \colon k[G] \to \mathrm{End}(M_k)$ (Proposition 1.5.9(ii)). It is defined, for every $f \in k[G]$, by $\Lambda(f) \colon m \mapsto fm$, where fm denotes the product in the module $_{k[G]}M$. Let $v_1, \ldots, v_n$ be a basis of M over k. Then $\Lambda(U(k[G])) \subseteq U(\mathrm{End}(M_k)) = \mathrm{GL}(M_k) \cong \mathrm{GL}_n(k)$ (this isomorphism depends on the fixed basis $v_1, \ldots, v_n$). Moreover, we have an embedding (i.e., an injective group morphism) of G in $U(k[G])$ defined by $g \mapsto 1_k \cdot g$, so that, if we restrict Λ to G, we get a representation ϱ of G.

Thus $\varrho(g)$ is the matrix associated to left multiplication by $1_k g \in k[G]$, which is an automorphism of M as a k-vector space, with respect to the basis $v_1, \ldots, v_n$.

If we fix a different basis $v'_1, \ldots, v'_n$ of M over k, then the resulting representation ϱ' of G turns out to be equivalent to ϱ (take as U the matrix of the change of basis.) More generally:

Proposition 3.6.3
Two representations of G over k are equivalent if and only if the corresponding modules over $k[G]$ are isomorphic.

Proof Let ϱ and ϱ' be two representations of G, and let $_{k[G]}k^n$ and $_{k[G]}k^m$ be the corresponding modules.

If the two modules are isomorphic via some $k[G]$-module isomorphism φ, then $n = m$, because they are isomorphic as k-vector spaces. Let U be the matrix associated to φ with respect to the canonical basis of $_k k^n$. Then, for any $f \in k[G]$ and any $\alpha := \begin{pmatrix} \alpha_1 \\ \vdots \\ \alpha_n \end{pmatrix} \in k^n$, $\varphi(f\alpha) = f\varphi(\alpha)$. In particular, for $f = 1_k g$, we have $\varphi(\varrho(g)\alpha) = \varrho'(g)\varphi(\alpha)$. Thus $U\varrho(g) = \varrho'(g)U$ for every $g \in G$. It follows that ϱ and ϱ' are equivalent.

Similarly for the inverse implication. □

Proposition 3.6.4
A representation ϱ of G is irreducible if and only if the corresponding $k[G]$-module is simple.

Proof Suppose ϱ irreducible of degree n. We must show that $_{k[G]}k^n$ is simple. We will assume $_{k[G]}k^n$ not simple and show that ϱ is reducible. If $_{k[G]}k^n$ is not simple, there exists a non-zero proper $k[G]$-submodule W of $_{k[G]}k^n$, which is *a fortiori* a non-zero proper k-vector subspace of k^n. Fix a basis $w_1, \ldots, w_m$ of W and complete it to a basis $w_1, \ldots, w_n$ of the whole space k^n. There is an isomorphism $\varphi: {}_{k[G]}k^n \to {}_k k^n$ of k-vector spaces that sends the canonical basis to the basis just specified. Via φ, we can endow the codomain of φ of a $k[G]$-module structure, in such a way that φ becomes a $k[G]$-module isomorphism. Hence the ϱ we started from will be equivalent to the representation associated to this new $k[G]$-module, the codomain of φ, by the previous proposition. The new representation, ϱ' say, sends g to $\alpha \mapsto 1_k g \alpha$. In particular, $\varrho'(g)(w_i)$ belongs to W for every $i = 1, 2, \ldots, m$,

3.6 Group Representations

so that the matrix of $\varrho'(g)$ with respect to the basis $w_1, \ldots, w_n$ is of the type $\begin{pmatrix} \varrho_1(g) & * \\ 0 & \varrho_2(g) \end{pmatrix}$. Thus ϱ is reducible.

Similarly for the inverse implication. □

> **Proposition 3.6.5**
> A representation ϱ of G is decomposable if and only if the corresponding $k[G]$-module is a direct sum of two non-zero proper submodules.

Proof Suppose that $_{k[G]}k^n$, the module associated to ϱ, is $= V \oplus W$ for some non-zero proper $k[G]$-submodules V, W. Fix k-bases $v_1, \ldots, v_r$ and $w_1, \ldots, w_s$ for V and W respectively, so that $r + s = n$. One has $V \cong {}_k k^r$ and $W \cong {}_k k^s$ as k-vector spaces, and the isomorphisms can carry over the $k[G]$-module structure, so that we have $V \cong {}_{k[G]}k^r$ and $W \cong {}_{k[G]}k^s$ and in the end $_{k[G]}k^n \cong {}_{k[G]}k^r \oplus {}_{k[G]}k^s$ as $k[G]$-modules. The representation ϱ' associated to the right-hand side sends g to $\alpha \mapsto 1_k g \alpha$; if α is in k^r, its image is in k^r as well, and if α is in k^s, its image is in k^s as well, hence the form of the matrix is $\begin{pmatrix} \varrho_1 & 0 \\ 0 & \varrho_2 \end{pmatrix}$, etc.

Similarly for the inverse implication. □

If we want to be more precise, it is possible to prove, for a fixed field k and group G, that we have an equivalence of categories between the full subcategory of $k[G]$–Mod, whose objects are the $k[G]$-modules of finite k-dimension and the category of all representations ρ of G over the field k (we haven't said what the morphisms in this category are, but it is easy to determine them as an exercise).

Now that we have set up our dictionary, we can present an application to group theory of our results about semisimple artinian rings.

> **Theorem 3.6.6** (Maschke, 1898)
> Let k be a field, G a finite group. Then $k[G]$ is a semisimple artinian ring if and only if $\operatorname{char} k$ does not divide $|G|$.

This means that $k[G]$ is semisimple artinian if and only if either $\operatorname{char} k = 0$, or $\operatorname{char} k = p$ and p does not divide $|G|$. For instance, $\mathbb{C}[G]$ is semisimple artinian for every finite group G. Thus all $\mathbb{C}[G]$-modules are semisimple, hence can be expressed uniquely as a direct sum of simples. From this, applied to the module associated to a given representation, we get that any representation ϱ over $\mathbb{C}$, up to equivalence, decomposes uniquely into irreducible representations over $\mathbb{C}$, i.e., ϱ is equivalent to

$$\begin{pmatrix} \varrho_1 & & 0 \\ & \ddots & \\ 0 & & \varrho_n \end{pmatrix}$$

with $\deg \varrho = \sum_i \deg \varrho_i$ and ϱ_i irreducible. Moreover, $k[G]$ has only finitely many simple modules up to isomorphism for every finite group G and every field k (because $k[G]$ is a finite-dimensional algebra, hence an artinian ring, and any artinian ring has only finitely many simple modules up to isomorphism.) Thus G has only finitely many irreducible representations over k up to equivalence.

We will prove a generalized form of Maschke's Theorem in Theorem 3.6.7. Notice that it is possible to define the ring $k[G]$ not only when G is a (finite) group and k is a field, but also for an arbitrary monoid G and an arbitrary ring k. If k is a ring and G is a monoid (written multiplicatively), that is, a semigroup with identity, then $k[G]$, called the *semigroup ring*, is the free left k-module with free set of generators G:

$$k[G] := \oplus_{g \in G} kg.$$

The multiplication is defined as in (3.5), the only difference is that now the λ_g's and the μ_h's are elements of the ring k, which are almost all zero. Then k is identified with $k \cdot 1_G$, which is a subring of $k[G]$, and G is identified with $1_k \cdot G$, which is a submonoid of the multiplicative monoid of $k[G]$. With these identifications, the elements of k commute with the elements of G. For instance, if k is a commutative ring, X is a set and G is the free monoid W_X, that is, the monoid of all words in the alphabet X, then $k[G]$ is the ring $k\langle X \rangle$ of polynomials with coefficients in k in the set of non-commuting indeterminates X. If k is a commutative ring and G is the commutative additive monoid $\mathbb{N}^n$, then $k[G] = k[\mathbb{N}^n]$ is the (usual) ring $k[x_1, \ldots, x_n]$ of polynomials with coefficients in k in the (commuting) indeterminates $x_1, \ldots, x_n$.

Theorem 3.6.7
Let k be an arbitrary ring and G be an arbitrary group. The ring $k[G]$ is semisimple artinian if and only if k is semisimple artinian, G is finite and $|G| \cdot 1_k$ is an invertible element of k.

Proof Let k be a semisimple artinian ring and G be a finite group of order $|G|$ with $|G| \cdot 1_k$ an invertible element of k. In order to show that $R := k[G]$ is a semisimple artinian ring, fix a right R-module M_R and an R-submodule N of M_R. It suffices to prove that N is a direct summand of M_R. As $k \subseteq R$, every right R-

3.6 Group Representations

module is a right k-module by restriction of scalars (Exercise 2.3.5). Thus we have the right module M_k and its k-submodule N, which is a direct summand of M_k because k is semisimple artinian. Hence we have a right k-module homomorphism $\pi: M \to N$ such that $\pi(y) = y$ for every $y \in N$. Let $u \in k$ be the inverse of $|G| \cdot 1_k$. Notice that u is in the center of k, hence in the center of R. For every $x \in M$, set $\varphi(x) := \sum_{g \in G} \pi(xg)g^{-1}u$. Then $\varphi(x) \in N$ for every $x \in M$, because $\pi(xg) \in N$ and N is an R-submodule of M. Thus we have defined a mapping $\varphi: M \to N$, which is clearly a right k-module homomorphism. In order to prove that it is also a right R-module homomorphism, fix $h \in G$. Then $\varphi(xh) = \sum_{g \in G} \pi(xhg)g^{-1}u$. Set $\ell := hg$, so that ℓ ranges in G when g ranges in G. Thus

$$\varphi(xh) = \sum_{\ell \in G} \pi(x\ell)(h^{-1}\ell)^{-1}u = \sum_{\ell \in G} \pi(x\ell)\ell^{-1}hu = \varphi(x)h.$$

It follows that φ is a right R-module homomorphism. Let $\varepsilon: N_R \to M_R$ be the embedding. For every $y \in N_R$, we have that

$$\varphi\varepsilon(y) = \sum_{g \in G} \pi(yg)g^{-1}u = \sum_{g \in G} yu = |G|yu = y.$$

Thus $\varphi\varepsilon$ is the identity morphism of N_R. By Lemma 2.5.3, N_R is a direct summand of M_R.

Conversely, assume $k[G]$ semisimple artinian. There is a ring homomorphism ε of $k[G]$ onto k, called the *augmentation map*, defined by

$$\varepsilon\Big(\sum_{g \in G} \lambda_g g\Big) = \sum_{g \in G} \lambda_g.$$

Thus k is a homomorphic image of $k[G]$, so that k must be semisimple artinian by Exercise 3.2.6(a).

The kernel of the augmentation map $\varepsilon: k[G] \to k$ is a two-sided ideal of $k[G]$, hence it is a fortiori a right ideal of $k[G]$. But $k[G]$ is semisimple artinian, so that $\ker \varepsilon$ must be a direct summand of $k[G]_{k[G]}$. Hence there exists a right ideal I of $k[G]$ with $I \oplus \ker \varepsilon = k[G]_{k[G]}$. By Proposition 2.9.3, there exists an idempotent $e \in k[G]$ with $I = ek[G]$ and $\ker \varepsilon = (1-e)k[G]$. Clearly, $1 - g \in \ker \varepsilon$ for every $g \in G$, so that $e(1-g) \in e \ker \varepsilon = e(1-e)k[G] = 0$. Thus $eg = e$ for every $g \in G$. Now $e = \sum_{g \in G} \lambda_g g$ for suitable $\lambda_g \in G$, almost all zero. The condition $eg = e$ for every $g \in G$ implies that all the λ_g's are equal. If G is infinite, then $\lambda_g = 0$ for all $g \in G$, so that $e = 0$ and $\varepsilon = 0$, a contradiction. This proves that the group G must be finite.

To conclude the proof, we must show that $|G| \cdot 1_k$ is an invertible element of k. Assume the contrary. Since $|G| \cdot 1_k$ is in the center of k, the principal right ideal $I := (|G| \cdot 1_k)k$ it generates is a proper two-sided ideal of k. Set $\overline{k} := k/I$, so that $\overline{k}$ is a homomorphic image of k, the group ring $\overline{k}[G]$ is a homomorphic image of $k[G]$

(hence it is semisimple artinian), and $|G| \cdot 1_{\bar{k}} = 0$. Set $\gamma := \sum_{g \in G} 1_{\bar{k}} g \in \bar{k}[G]$. Then γ is a non-zero element and $h\gamma = \gamma h = \gamma$ for all $h \in G$ because multiplication by h just permutes the summands but does not change the sum. This implies that γ is a central element of $\bar{k}[G]$ and $\gamma^2 = \sum_{g \in G} g\gamma = \sum_{g \in G} \gamma = |G|\gamma = (|G| \cdot 1_{\bar{k}})\gamma = 0$. Thus $\gamma \bar{k}[G]$ is a non-zero nilpotent two-sided ideal of $\bar{k}[G]$. This is a contradiction, because $\bar{k}[G]$ is semisimple artinian. □

4 Local Rings, Injective Modules, Flat Modules

4.1 Local Rings

In this Section, we will introduce an important class of rings, the class of local rings.

Proposition 4.1.1
The following conditions are equivalent for a ring $R \neq 0$:

(i) *The ring R has a unique maximal right ideal.*
(ii) *The Jacobson radical $J(R)$ is a maximal right ideal.*
(iii) *The sum of two elements of R that are not right invertible is not right invertible.*
(iv) $J(R) = \{ r \in R \mid rR \neq R \}$.
(v) $R/J(R)$ *is a division ring.*
(vi) $J(R) = \{ r \in R \mid r \text{ is not invertible in } R \}$.
(vii) *The sum of two non-invertible elements of R is non-invertible.*
(viii) *For every $r \in R$, either r is invertible or $1 - r$ is invertible.*

Notice that some of these conditions are right/left symmetric, so that right can be replaced with left in the other conditions. The rings that satisfy the equivalent conditions of Proposition 4.1.1 are called *local rings*. By definition, the zero ring is not a local ring. (The reader can check that some of the properties (i)–(viii) hold for the zero ring and some do not.)

Proof

(i) $\Rightarrow$ (ii)   If the ring R has a unique maximal right ideal M, then $J(R) = M$ because $J(R)$ is the intersection of all maximal right ideals of R. Hence $J(R)$ is a maximal right ideal.

(ii) $\Rightarrow$ (i)   Since $J(R)$ is a maximal right ideal and is the intersection of all maximal right ideals of R, the maximal right ideal $J(R)$ must be contained in every maximal right ideal of R, so that $J(R)$ is the unique maximal right ideal of R.

(i) $\Rightarrow$ (iii)   Assume that R has a unique maximal right ideal M. Notice that an element r of a ring R is not right invertible if and only if $1 \notin rR$, that is, if and only if the right ideal rR of R is proper. Since every proper right ideal of R is contained in a maximal right ideal, it follows that M contains all proper right ideals. Thus if r, s are not right invertible, then $(r+s)R \subseteq rR + sR \subseteq M \subset R$. Hence $r+s$ is not right invertible.

(iii) $\Rightarrow$ (iv)   Suppose (iii) holds. Since $J(R)$ is a proper ideal, for every $r \in J(R)$, the right ideal rR is strictly contained in R, so $rR \neq R$. Conversely if $x \in R$ and $xR \neq R$, then, for every $r \in R$, $xrR \neq R$, so xr is not right invertible. But $xr + (1 - xr) = 1$. By (iii), $1 - xr$ is right invertible. Thus $x \in J(R)$ by Proposition 3.5.3(ii).

(iv) $\Rightarrow$ (v)   Take $0 \neq \overline{x} \in R/J(R)$. Then $x \in R$ and $x \notin J(R)$, so by (iv), $xR = R$, so x is right invertible, so $\overline{x}$ is right invertible.

(v) $\Rightarrow$ (vi)   Suppose that $R/J(R)$ is a division ring. Then clearly every element of $J(R)$ is non invertible. Conversely, suppose that $x \in R$ and $x \notin J(R)$. Then $\overline{x} \neq 0$ in $R/J(R)$, so it is invertible in $R/J(R)$, i.e. there exists $y \in R$ such that $xy - 1, yx - 1 \in J(R)$. Hence xy, yx are invertible in R. Thus x is invertible.

(vi) $\Rightarrow$ (vii)   is trivial, because $J(R)$ is closed under addition.

(vii) $\Rightarrow$ (viii)   Assume that (viii) does not hold, so that there exists an $r \in R$ such that both r and $1-r$ are non-invertible. Since $r + (1-r) = 1$ is invertible, (vii) does not hold.

(viii) $\Rightarrow$ (vi)   Suppose that (viii) holds. We will prove that the equality in (vi) holds. The inclusion $\subseteq$ is obvious, because if $r \in J(R)$, then $rR \subseteq J(R) \subset R$, so that x in not right invertible. For $\supseteq$, fix a non-invertible element $r \in R$. If r is not right invertible, then rx is not right invertible for every $x \in R$. By (viii), the element $1 - rx$ is invertible for every $x \in R$. Thus $r \in J(R)$. Similarly, if r is not left invertible, then $1 - xr$ is invertible for every $x \in R$, so that $r \in J(R)$.

(vi) $\Rightarrow$ (v)   is easy.

(v) $\Rightarrow$ (ii)   If $R/J(R)$ is a division ring, $R/J(R)$ has only the trivial right ideals. In view of the Correspondence Theorem between right ideals of $R/J(R)$ and right ideals of R containing $J(R)$, the unique right ideals of R containing $J(R)$ are R and $J(R)$. But all maximal right ideals of R contain $J(R)$, so that $J(R)$ is the unique maximal right ideal of R.

$\square$

4.1 Local Rings

Proposition 4.1.2
Local rings have IBN.

Proof If R is local, $R/J(R)$ is a division ring, hence $R/J(R)$ has IBN. Thus R also has IBN by Exercise 2.4.1(4). □

We will not prove that:

Theorem 4.1.3 (Kaplansky [16, Theorem 2])
Every projective module over a local ring is free.

Exercise 4.1.4
Show that if $p \in \mathbb{Z}$ is prime and $\mathbb{Z}_{(p)}$ is the set of all rational numbers that can be written in the form x/y, where $x, y \in \mathbb{Z}$ and p does not divide y, then $\mathbb{Z}_{(p)}$ is a commutative local ring whose maximal ideal is the principal ideal generated by p. Notice that the rings $\mathbb{Z}_{(p)}$ are subrings of $\mathbb{Q}$ that contain $\mathbb{Z}$.

Exercise 4.1.5
Let Γ be the semigroup consisting of all the symbols x^q, where $q \in \mathbb{Q}$, $q \geq 0$, and the multiplication is defined by $x^q x^{q'} := x^{q+q'}$. Let K be a field. Form the semigroup algebra $R := K[\Gamma]$, and set $\overline{R} := R/x^1 R$. Show that $\overline{R}$ is a local ring with Jacobson radical $\overline{J} := J(\overline{R})$, a nil but not nilpotent ideal. Show that $\overline{J}^2 = \overline{J} \neq 0$, so that Nakayama's lemma can fail for infinitely generated modules.

Exercise 4.1.6
Show that if R is a ring and the lattice $\mathcal{L}(R_R)$ of all its right ideals is linearly ordered under $\subseteq$, then R is a local ring. (A ring R for which the lattice $\mathcal{L}(R_R)$ of all right ideals is linearly ordered is called a right *chain ring*.)

A module M_R is *indecomposable* if $M_R \neq 0$ and whenever $M_R = A \oplus B$, with A, B submodules of M_R, then either $A = 0$ or $B = 0$. For instance, simple modules are indecomposable. The $\mathbb{Z}$-module $\mathbb{Z}$ is not simple, but it is indecomposable, because the intersection of any two non-zero submodules of $\mathbb{Z}$ is non-zero.

Proposition 4.1.7
Let M_R be a module over an arbitrary ring R and assume that $\mathrm{End}(M_R)$ is a local ring. Then the right R-module M_R is indecomposable.

Proof Assume that M_R is not indecomposable. If $M_R = 0$, then $\text{End}(M_R) = 0$ is not a local ring. Thus $M_R \neq 0$ and $M_R = A \oplus B$, with A, B non-zero submodules of M_R. There is an endomorphism e of M_R that is the identity on A and zero on B. Then $1 - e$ is an endomorphism of M_R that is zero on A and the identity on B. Since $A \neq 0$ and $B \neq 0$, both e and $1 - e$ have a non-zero kernel, hence they are not invertible elements of the ring $\text{End}(M_R)$. Thus $\text{End}(M_R)$ is not local by Proposition 4.1.1(viii). □

Lemma 4.1.8
Let M be a module and f an endomorphism of M.

(a) *If n is a positive integer such that $f^n(M) = f^{n+1}(M)$, then*

$$\ker(f^n) + f^n(M) = M.$$

(b) *If M is an artinian module, then f is an automorphism if and only if f is injective.*

Proof
(a) If n is such that $f^n(M) = f^{n+1}(M)$, then $f^t(M) = f^{t+1}(M)$ for every $t \geq n$, so that $f^n(M) = f^{2n}(M)$. Let us show that

$$\ker(f^n) + f^n(M) = M.$$

If $x \in M$, then $f^n(x) \in f^n(M) = f^{2n}(M)$, so that $f^n(x) = f^n(y)$ for some $y \in f^n(M)$. Therefore $x - y \in \ker(f^n)$, and $x = (x - y) + y \in \ker(f^n) + f^n(M)$.

(b) If f an injective endomorphism of the artinian module M, the descending chain

$$f(M) \supseteq f^2(M) \supseteq f^3(M) \supseteq \ldots$$

is stationary, so that $\ker(f^n) + f^n(M) = M$ for some positive integer n by part (a). As f^n is injective, $\ker(f^n) = 0$, and therefore $f^n(M) = M$. In particular, f is surjective. □

4.1 Local Rings

Similarly it can be proved that

Lemma 4.1.9
Let M be a module and f an endomorphism of M.

(a) *If n is a positive integer such that $\ker f^n = \ker f^{n+1}$, then*
$$\ker(f^n) \cap f^n(M) = 0.$$

(b) *If M is a noetherian module, then f is an automorphism if and only if f is surjective.*

Lemma 4.1.10 (Fitting's Lemma)
Let M be a module of finite composition length n. Then:

(a) $M = \ker(f^n) \oplus f^n(M)$ *for every endomorphism f of M.*
(b) *If M is indecomposable, the ring $\mathrm{End}(M_R)$ is local.*

Proof
(a) Since M is of finite composition length n, in the descending chain
$$M \supseteq f(M) \supseteq f^2(M) \supseteq f^3(M) \supseteq \cdots \supseteq f^{n+1}(M),$$
which is of length $n + 1$, not all the inclusions can be proper. Therefore there exists $m \le n$ such that $f^m(M) = f^{m+1}(M)$. It follows that $f^n(M) = f^{n+1}(M)$. Similarly, considering the chain
$$0 \subseteq \ker f \subseteq \ker f^2 \subseteq \ker f^3 \subseteq \cdots \subseteq \ker f^{n+1},$$
we get that $\ker f^n = \ker f^{n+1}$. Thus $\ker(f^n) \oplus f^n(M) = M$ by Lemmas 4.1.8(a) and 4.1.9(a).

(b) Assume M_R indecomposable. It suffices to show that for every $f \in \mathrm{End}(M_R)$, either f is invertible or $1 - f$ is invertible. Fix $f \in \mathrm{End}(M_R)$. By (a), $M = \ker(f^n) \oplus f^n(M)$. As M is indecomposable, two cases may occur. In the first case, $\ker(f^n) = 0$ and $f^n(M) = M$. Then f^n is an automorphism of M, so that f is an automorphism of M, i.e., f is invertible in $\mathrm{End}(M_R)$. In the second case, $\ker(f^n) = M$, that is, f is a nilpotent element of the ring $\mathrm{End}(M_R)$. It follows that $1 - f$ is invertible in the ring $\mathrm{End}(M)$. $\square$

We conclude this Section with four important results about direct-sum decompositions of modules.

> **Proposition 4.1.11**
> *If a module M is a direct sum of modules with local endomorphism rings, then every indecomposable direct summand of M has local endomorphism ring.*

We don't give a proof of Proposition 4.1.11. The interested reader can find a proof in [9, Lemma 2.11].

> **Lemma 4.1.12**
> *Let M be a module with two direct-sum decompositions*
> $$M = \oplus_{i \in I} M_i = N \oplus N'.$$
> *Suppose that all the modules M_i ($i \in I$) and the module N have local endomorphism rings. Then there exists an index $i \in I$ such that:*
>
> (a) *the composite mapping of the inclusion $\varepsilon_i : M_i \to M$ and the projection $\pi_N : M \to N$ with kernel N' is an isomorphism of M_i onto N.*
> (b) $M = M_i \oplus N'$.

Proof The module N has a local endomorphism ring, hence a non-zero endomorphism ring, so $N \neq 0$ contains a non-zero element $n \in N$. Write $n = m_1 + m_2 + \cdots + m_k$, where, for every $s = 1, 2, \ldots, k$, m_s is a non-zero element of M_{i_s}. Set $P = M_{i_1} \oplus M_{i_2} \oplus \cdots \oplus M_{i_k}$. Let Q be the direct sum of all the modules M_i where i ranges in $I \setminus \{i_1, i_2, \ldots, i_k\}$, so that $M = P \oplus Q$. Let $\varepsilon_N : N \to M$, $\varepsilon_P : P \to M$, $\varepsilon_Q : Q \to M$ be the inclusions, and $\pi_N : M \to N$, $\pi_P : M \to P$, $\pi_Q : M \to Q$ denote the projections with kernel N', Q, P, respectively. Set $\varphi := \pi_N \varepsilon_P \pi_P \varepsilon_N$ and $\psi := \pi_N \varepsilon_Q \pi_Q \varepsilon_N$. Then φ and ψ are endomorphisms of N, and

$$\varphi + \psi = \pi_N \varepsilon_P \pi_P \varepsilon_N + \pi_N \varepsilon_Q \pi_Q \varepsilon_N = \pi_N (\varepsilon_P \pi_P + \varepsilon_Q \pi_Q) \varepsilon_N$$
$$= \pi_N \iota_M \varepsilon_N = \pi_N \varepsilon_N = \iota_N.$$

Since the endomorphism ring of N is local, we have that either φ or ψ is an automorphism of N. But $\psi(n) = 0$, so that φ must be an automorphism.

Now let $\varepsilon_s : M_{i_s} \to M$ and $\pi_s : M \to M_{i_s}$ denote the inclusion and the projection for every $s = 1, 2, \ldots, k$, and consider the composite mappings $\varphi_s := \pi_N \varepsilon_s \pi_s \varepsilon_N$. Then $\varphi_1 + \cdots + \varphi_k = \varphi$, so that there exists $s = 1, 2, \ldots, k$ such that

4.1 Local Rings

φ_s is an automorphism of N. Thus $(\varphi_s)^{-1}\pi_N\varepsilon_s\pi_s\varepsilon_N$ is the identity automorphism of N. Applying Lemma 2.5.3 to the homomorphisms

$$f := \pi_s\varepsilon_N: N \to M_{i_s} \quad \text{and} \quad g := (\varphi_s)^{-1}\pi_N\varepsilon_s: M_{i_s} \to N,$$

so $gf = \iota_N$, we get that $M_{i_s} = f(N) \oplus \ker g$, g is an epimorphism, and f is a monomorphism. Thus $f(N) \neq 0$ and $\ker g = 0$. Therefore g is an isomorphism, so $\pi_N\varepsilon_s: M_{i_s} \to N$ is an isomorphism as well. This proves (a).

Now $(\pi_N\varepsilon_s)^{-1}\pi_N\varepsilon_s$ is the identity automorphism of M_{i_s}. Applying Lemma 2.5.3 to the homomorphisms $\varepsilon_s: M_{i_s} \to M$ and $(\pi_N\varepsilon_s)^{-1}\pi_N: M \to M_{i_s}$, we get that $M = \varepsilon_s(M_{i_s}) \oplus \ker((\pi_N\varepsilon_s)^{-1}\pi_N) = M_{i_s} \oplus N'$, which concludes the proof of the Lemma. □

Theorem 4.1.13 (Krull-Schmidt-Remak-Azumaya Theorem)
Let M be a module that is a direct sum of modules with local endomorphism rings. Then any two direct-sum decompositions of M into indecomposable direct summands are isomorphic. That is, if M is a direct sum of modules with local endomorphism rings and $M = \oplus_{i \in I} M_i = \oplus_{j \in J} N_j$, where the M_i, N_j are indecomposable modules, then there exists a one-to-one correspondence $\varphi: I \to J$ such that $M_i \cong N_{\varphi(i)}$ for every $i \in I$.

Proof Suppose $M = \oplus_{i \in I} M_i = \oplus_{j \in J} N_j$, where M_i has a local endomorphism ring for every $i \in I$ and N_j is indecomposable for all $j \in J$. By Proposition 4.1.11, all modules N_j also have a local endomorphism ring. For each $i \in I$ define a subset $I(i)$ of I, consisting of all indices $j \in I$ such that $M_i \cong M_j$. The subsets $I(i)$ clearly form a partition of I. Similarly, for each $i \in I$, define a subset $J(i)$ of J, consisting of all indices $j \in J$ such that $M_i \cong N_j$. The subsets $J(i)$ also form a partition of J by Lemma 4.1.12.

Let us show that for every $i \in I$ the cardinality of $J(i)$ is less or equal to the cardinality of $I(i)$. Consider an element $j_1 \in J(i)$. By Lemma 4.1.12, there exists an index $i'_1 \in I(i)$ such that $M = M_{i'_1} \oplus (\oplus_{k \neq j_1} N_k)$. Now fix another element $j_2 \in J(i)$, $j_1 \neq j_2$. Consider the splitting epimorphism $\pi: M \to M/M_{i'_1}$. Applying π to the equalities $M = \oplus_{l \in I} M_l = M_{i'_1} \oplus (\oplus_{k \neq j_1} N_k)$, we find that $M/M_{i'_1} = \pi(M) = \oplus_{l \neq i'_1} \pi(M_l) = \oplus_{k \neq j_1} \pi(N_k)$. All these modules $\pi(M_l)$, $\pi(N_k)$, ($l \in I, l \neq i'_1, k \in J, k \neq j_1$) are isomorphic to M_l, N_k, respectively, hence they all have a local endomorphism ring. By Lemma 4.1.12, there exists an index $i'_2 \in I(i) \setminus \{i'_1\}$ such that $\pi(M) = \pi(M_{i'_2}) \oplus (\oplus_{k \neq j_1, j_2} \pi(N_k))$. Taking the inverse image of this equality via the splitting epimorphism π, we get that $M = M_{i'_1} \oplus M_{i'_2} \oplus (\oplus_{k \neq j_1, j_2} N_k)$. Repeating recursively this construction, we see that for every finite set of distinct elements in $J(i)$ there is a finite set of the same cardinality of distinct elements in $I(i)$.

Now the roles of the modules M_i and N_j can be inverted, so that for every finite set of distinct elements in $I(i)$ there is a finite set of the same cardinality of distinct elements in $J(i)$. As a consequence we get that the set $I(i)$ is finite if and only if $J(i)$ is finite, and in this case $I(i)$ and $J(i)$ have the same cardinality.

We will now consider the remaining case of $I(i)$ and $J(i)$ both infinite. Fix an index $a \in I(i)$. For every $b \in J$ we have a homomorphism $\pi_b \varepsilon_a \colon M_a \to N_b$. Let $K(a)$ be the set of all $b \in J$ such that $\pi_b \varepsilon_a$ is an isomorphism. Then $K(a) \subseteq J(i)$. Let m be a non-zero element of M_a. Write $m = n_1 + n_2 + \cdots + n_k$, where, for every $s = 1, 2, \ldots, k$, n_s is a non-zero element of N_{j_s}. Then $\pi_b \varepsilon_a(m) = \pi_b(m) = 0$ for every $b \in J \setminus \{j_1, \ldots, j_k\}$. Therefore $K(a)$ is a finite set for every $a \in I(i)$. Also $\bigcup_{a \in I(i)} K(a) = J(i)$, because by Lemma 4.1.12(a), for each element $b \in J(i)$ there exists an index $a \in I(i)$ such that $\pi_b \varepsilon_a \colon M_a \to N_b$ is an isomorphism. Considering cardinalities, we get that

$$|J(i)| = \left| \bigcup_{a \in I(i)} K(a) \right| \leq \sum_{a \in I(i)} |K(a)| \leq |I(i)| \aleph_0 = |I(i)|.$$

Inverting the roles of the modules M_i and N_j, we find that $|I(i)| \leq |J(i)|$. By the Cantor-Schröder-Bernstein Theorem (Theorem 6.2.2), if follows that $|I(i)| = |J(i)|$, so that for each $i \in I$ there is a bijection $\varphi_i \colon I(i) \to J(i)$. Gluing together all these bijections φ_i, we get a one-to-one correspondence $\varphi \colon I \to J$ with the property required in the statement of the Theorem. □

Corollary 4.1.14 (Krull-Schmidt)
Let M be a module of finite composition length. Then any two direct-sum decompositions of M into indecomposable direct summands are isomorphic. That is, if $M = M_1 \oplus \cdots \oplus M_n = N_1 \oplus \cdots \oplus N_m$, where $M_1, \ldots, M_n, N_1, \ldots, N_m$ are indecomposable modules, and M has finite composition length, then $n = m$ and there exists a permutation σ of $\{1, 2, \ldots, n\}$ such that $M_i \cong N_{\sigma(i)}$ for every $i = 1, 2, \ldots, n$.

The proof now easily follows from Fitting's Lemma 4.1.10 and the Krull-Schmidt-Remak-Azumaya Theorem 4.1.13.

4.2 Injective Modules

Fix two modules M_R and N_R. We already know that there are a covariant functor

$$\mathrm{Hom}(M_R, -) \colon \mathrm{Mod}{-}R \to \mathbf{Ab}$$

4.2 Injective Modules

and a contravariant functor

$$\mathrm{Hom}(-, N_R) \colon \mathrm{Mod}{-}R \to \mathbf{Ab}.$$

Also, we have already seen in Exercise 2.7.12 that these functors Hom are "left exact", in the sense that, for every fixed module M_R, if $0 \to N'_R \to N_R \to N''_R$ is exact, then so is $0 \to \mathrm{Hom}(M_R, N'_R) \to \mathrm{Hom}(M_R, N_R) \to \mathrm{Hom}(M_R, N''_R)$, and, for every fixed module N_R, if $M'_R \to M_R \to M''_R \to 0$ is exact, then so is $0 \to \mathrm{Hom}(M''_R, N_R) \to \mathrm{Hom}(M_R, N_R) \to \mathrm{Hom}(M'_R, N_R)$.

In general, these functors $\mathrm{Hom}(M_R, -)$ and $\mathrm{Hom}(-, N_R)$ are not "exact", that is, it is not always true that, for every fixed module M_R, if $0 \to N'_R \to N_R \to N''_R \to 0$ is a short exact sequence, then $0 \to \mathrm{Hom}(M_R, N'_R) \to \mathrm{Hom}(M_R, N_R) \to \mathrm{Hom}(M_R, N''_R) \to 0$ is necessarily exact, and, for every fixed module N_R, if $0 \to M'_R \to M_R \to M''_R \to 0$ is exact, then $0 \to \mathrm{Hom}(M''_R, N_R) \to \mathrm{Hom}(M_R, N_R) \to \mathrm{Hom}(M'_R, N_R) \to 0$ is necessarily exact. It is easily seen that a module M_R is projective if and only if the functor $\mathrm{Hom}(M_R, -)$ is exact, that is, for every exact sequence $0 \to N'_R \to N_R \to N''_R \to 0$, the sequence of abelian groups $0 \to \mathrm{Hom}(M_R, N'_R) \to \mathrm{Hom}(M_R, N_R) \to \mathrm{Hom}(M_R, N''_R) \to 0$ is exact.

We leave to the reader the proof of the following easy result. It is just the dual of the corresponding result that holds for projective modules.

> **Proposition 4.2.1**
> *The following conditions are equivalent for an R-module E_R:*
>
> (i) *The functor $\mathrm{Hom}(-, E_R) \colon \mathrm{Mod}{-}R \to \mathbf{Ab}$ is exact, that is, for every exact sequence $0 \to M'_R \to M_R \to M''_R \to 0$ of right R-modules, the sequence of abelian groups $0 \to \mathrm{Hom}(M''_R, E_R) \to \mathrm{Hom}(M_R, E_R) \to \mathrm{Hom}(M'_R, E_R) \to 0$ is exact.*
> (ii) *For every monomorphism $M'_R \to M_R$ of right R-modules,*
>
> $$\mathrm{Hom}(M_R, E_R) \to \mathrm{Hom}(M'_R, E_R)$$
>
> *is an epimorphism of abelian groups.*
> (iii) *For every submodule M'_R of a right R-module M_R, every morphism $M'_R \to E_R$ can be extended to a morphism $M_R \to E_R$.*
> (iv) *For every monomorphism $f \colon M'_R \to M_R$ and every homomorphism $g \colon M'_R \to E_R$, there exists a morphism $h \colon M_R \to E_R$ with $h \circ f = g$.*

A module E_R is *injective* if it satisfies the equivalent conditions of Proposition 4.2.1.

Condition (iv) is described by the following commutative diagram, in which the row is exact:

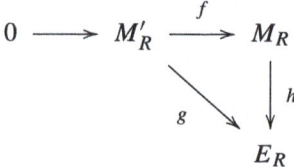

Essentially, the unique characterization of projective modules that cannot be immediately dualized to injective modules is the characterization of projective modules as direct summands of free modules. In the next Proposition, we give a further criterion to determine injective modules, that is, a further characterization of injective modules, which does not have an analog for projective modules.

Proposition 4.2.2 (Baer's Criterion)
A right module E over a ring R is injective if and only if for every right ideal I of R, every right R-module morphism $\sigma\colon I \to E$ can be extended to a morphism $\sigma^*\colon R_R \to E_R$.

Proof ($\Rightarrow$) is trivial. ($\Leftarrow$): Assume that every morphism $I \to E$ extends to a morphism $R_R \to E$. Fix $M' \le M_R$ and a morphism $\varphi\colon M' \to E$. Consider the set

$$\mathcal{F} := \{(N, \psi) \mid M' \le N \le M,\ \psi\colon N \to E \text{ extends } \varphi\}.$$

Then $\mathcal{F} \ne \emptyset$, because $(M', \varphi) \in \mathcal{F}$. Partially order $\mathcal{F}$ setting $(N, \psi) \le (N', \psi')$ if $N \subseteq N'$ and $\psi'|_N = \psi$. It is easily seen that $\le$ is a partial order on $\mathcal{F}$. We claim that every chain in $\mathcal{F}$ has an upper bound in $\mathcal{F}$. In order to prove the claim, fix a chain $\{(N_\lambda, \psi_\lambda) \mid \lambda \in \Lambda\}$ in $\mathcal{F}$. Then $\{N_\lambda \mid \lambda \in \Lambda\}$ is a chain of submodules of M_R containing M', so $\bigcup_\lambda N_\lambda$ is a submodule of M, which clearly contains M'. Define $\overline{\psi}\colon \bigcup_\lambda N_\lambda \to E$ as follows. If $x \in \bigcup_\lambda N_\lambda$ there exists $\overline{\lambda}$ with $x \in N_{\overline{\lambda}}$. Set $\overline{\psi}(x) := \psi_{\overline{\lambda}}(x)$. It is easily seen that this function is a well defined morphism and that $(\bigcup_\lambda N_\lambda, \overline{\psi})$ is an upper bound for the chain $(N_\lambda, \psi_\lambda)$, $\lambda \in \Lambda$. This proves the claim. Applying Zorn's lemma, we find that there exists a maximal element $(\overline{N}, \overline{\psi}) \in \mathcal{F}$. We will now show that $\overline{N} = M$.

Assume by contradiction that $\overline{N} \ne M$, and fix $b \in M \setminus \overline{N}$. Let $I := \{r \in R \mid br \in \overline{N}\} \le R_R$. Let $\sigma\colon I_R \to E_R$ be defined by $\sigma(r) := \overline{\psi}(br)$ for every $r \in I$. By

4.2 Injective Modules

hypothesis, there is a morphisms $\sigma^*: R_R \to E_R$ that extends σ. Consider now

$$\psi: \overline{N} + bR \to E_R$$

$$c + br \mapsto \overline{\psi}(c) + \sigma^*(r).$$

The mapping ψ is well defined, because if $c + br = c' + br'$, then $r - r' \in I$, so that

$$\sigma^*(r) - \sigma^*(r') = \sigma^*(r - r') = \overline{\psi}(b(r - r')) = \overline{\psi}(c' - c) = \overline{\psi}(c') - \overline{\psi}(c).$$

One easily sees that ψ is a right R-module morphism that extends φ, which contradicts the maximality of $(\overline{N}, \overline{\psi})$.

Therefore $\overline{N} = M$, and we are done. $\square$

Now we are going to determine injective $\mathbb{Z}$-modules, that is, injective abelian groups.

> **Definition 4.2.3**
>
> An additive abelian group G is *divisible* if $nG = G$ for every non-zero integer n (equivalently, for every positive integer n). Thus G is divisible if and only if, for every $g \in G$ and every $n > 0$, there exists $h \in G$ such that $nh = g$.

For instance, the abelian group $\mathbb{Z}$ is not divisible, and the abelian group $\mathbb{Q}$ is divisible. Homomorphic images of divisible abelian groups are divisible. It would be possible to prove that every divisible abelian group is a direct sum of copies of $\mathbb{Q}$ and Prüfer groups $\mathbb{Z}(p^\infty)$.

> **Proposition 4.2.4**
>
> A $\mathbb{Z}$-module G is injective if and only if it is a divisible abelian group.

Proof Let G be an injective $\mathbb{Z}$-module. If $g \in G$ and $n > 0$, there is a $\mathbb{Z}$-module homomorphism $\varphi: n\mathbb{Z} \to G$ defined by $\varphi(nz) = zg$ for every $z \in \mathbb{Z}$. As G is injective, φ extends to a $\mathbb{Z}$-module morphism $\psi: \mathbb{Z} \to G$. Then $h := \psi(1)$ is an element of G such that $nh = n\psi(1) = \psi(n) = \varphi(n) = g$. This proves that G is divisible.

Conversely, let G be a divisible abelian group. In order to prove that the module $G_\mathbb{Z}$ is injective, by Baer's criterion it suffices to show that for every ideal I of $\mathbb{Z}$, every morphism $\varphi: I \to G$ extends to $\mathbb{Z}$. Now every ideal I of $\mathbb{Z}$ is of the form $n\mathbb{Z}$ with $n \geq 0$. Clearly, if $n = 0$ every morphism $\varphi: n\mathbb{Z} \to G$ extends to $\mathbb{Z}$. Assume $n > 0$, and let $\varphi: n\mathbb{Z} \to G$ be a homomorphism. Since G is divisible, there exists

$h \in G$ such that $nh = \varphi(n)$. Let $\psi\colon \mathbb{Z} \to G$ be the $\mathbb{Z}$-module homomorphism defined by $\psi(z) = zh$ for every $z \in \mathbb{Z}$. Then ψ extends φ, because, for every $z \in \mathbb{Z}$, $\varphi(nz) = z\varphi(n) = z(nh) = (nz)h = \psi(nz)$. Thus $G_{\mathbb{Z}}$ is injective. □

Exercise 4.2.5
Show that an abelian group is divisible if and only if it is a homomorphic image of $\mathbb{Q}^{(X)}$ for some set X.

> **Proposition 4.2.6**
> *Direct summands of injective modules are injective.*

Proof Assume E_R injective. Let F_R be a direct summand of E_R. Fix modules $A_R \le B_R$ and a morphism $\varphi\colon A_R \to F_R$. We must show that φ extends to B_R. Let $\varepsilon\colon F_R \hookrightarrow E_R$ and $\pi\colon E_R \to F_R$ be the inclusion and the canonical projection, so that $\pi\varepsilon = 1_F$. Let $i\colon A_R \hookrightarrow B_R$ be the inclusion. Since E_R is injective, the morphism $\varepsilon\varphi\colon A_R \to E_R$ extends to B_R, that is, there is a morphism $\varphi'\colon B_R \to E_R$ with $\varphi'i = \varepsilon\varphi$. Then $\pi\varphi'\colon B_R \to F_R$ extends $\varphi\colon A_R \to F_R$, because $\pi\varphi'i = \pi\varepsilon\varphi = \varphi$. □

> **Proposition 4.2.7**
> *Direct products of injective modules are injective modules.*

Proof Let $(E_\lambda)_{\lambda \in \Lambda}$ be a family of injective modules, A_R be a submodule of B_R, and $\varphi\colon A \to \prod_{\lambda \in \Lambda} E_\lambda$ be a homomorphism. Let $\pi_\mu\colon \prod_{\lambda \in \Lambda} E_\lambda \to E_\mu$ be the canonical projection for every $\mu \in \Lambda$. Since E_μ is injective, $\pi_\mu\varphi\colon A \to E_\mu$ extends to a morphism $\psi_\mu\colon B \to E_\mu$, for every $\mu \in \Lambda$. There exists a unique morphism $\psi\colon B \to \prod_{\lambda \in \Lambda} E_\lambda$ such that $\psi_\mu = \pi_\mu\psi$ for every $\mu \in \Lambda$. It is easily seen that ψ extends φ. □

In particular, a direct sum of finitely many injective modules is an injective module.

It would be possible to prove (Bass-Papp Theorem) that any direct sum of injective right R-modules is injective if and only if R is a right Noetherian ring.

In the next proposition we consider the functor $\mathrm{Hom}(_R R_{\mathbb{Z}}, -)\colon \mathbf{Ab} \to \mathrm{Mod}{-}R$.

4.2 Injective Modules

Proposition 4.2.8
If R is a ring and G is a divisible abelian group, then

$$\mathrm{Hom}(R_{\mathbb{Z}}, G_{\mathbb{Z}})$$

is an injective right R-module.

Here the right R-module structure on $\mathrm{Hom}(R_{\mathbb{Z}}, G_{\mathbb{Z}})$ is that induced by the bimodule structure on ${}_R R_{\mathbb{Z}}$. Hence, for every $f \in \mathrm{Hom}(R_{\mathbb{Z}}, G_{\mathbb{Z}})$ and every $r \in R$, it is defined by $fr := f \circ \lambda_r$, where $\lambda_r \colon R \to R$ denotes left multiplication by r.

Proof We will apply Baer's criterion. Fix a right ideal I of R, and a right R-module morphism $\varphi \colon I \to \mathrm{Hom}(R_{\mathbb{Z}}, G_{\mathbb{Z}})$. We must show that φ extends to R_R. Now the $\mathbb{Z}$-module morphism $\psi \colon I \to G$, defined by $\psi(a) = \varphi(a)(1_R)$, extends to a $\mathbb{Z}$-module morphism $\psi^* \colon R_{\mathbb{Z}} \to G_{\mathbb{Z}}$, because G is divisible, i.e., injective as a $\mathbb{Z}$-module (Proposition 4.2.4). Let $\varphi^* \colon R_R \to \mathrm{Hom}(R_{\mathbb{Z}}, G_{\mathbb{Z}})$ be the right R-module morphism such that $\varphi^*(1_R) = \psi^*$. It remains to show that φ^* extends φ. Fix $a \in I$. Then, for every $r \in R$, we have

$$\varphi^*(a)(r) = (\psi^* \circ \lambda_a)(r) = \psi^*(ar) = \psi(ar) =$$
$$= \varphi(ar)(1_R) = \varphi(a)(\lambda_r(1_R)) = \varphi(a)(r),$$

that is, $\varphi^*(a) = \varphi(a)$ for every $a \in I$. $\square$

Theorem 4.2.9
Every right R-module can be embedded in an injective right R-module.

Proof Suppose first that $R = \mathbb{Z}$. If G is a $\mathbb{Z}$-module, that is, an abelian group, then G is a homomorphic image of a free abelian group, so that there exists a set X and a subgroup H of the free abelian group $\mathbb{Z}^{(X)}$ with $G \cong \mathbb{Z}^{(X)}/H$. Hence G is isomorphic to a subgroup of $\mathbb{Q}^{(X)}/H$, which is a divisible group because it is a homomorphic image of the divisible abelian group $\mathbb{Q}^{(X)}$. This proves the theorem in the case $R = \mathbb{Z}$.

Now let R be an arbitrary ring and M_R a right R-module. We know that $M_{\mathbb{Z}} \subseteq G$ for a suitable divisible abelian group G, as we have seen in the previous paragraph. Set $\varepsilon \colon M_R \to \mathrm{Hom}(R_{\mathbb{Z}}, G_{\mathbb{Z}})_R$, $\varepsilon(x)(r) = xr$ for every $x \in M_R, r \in R$. It is easily seen that ε is a right R-module monomorphism. As $\mathrm{Hom}_{\mathbb{Z}}(R, G)$ is injective by Proposition 4.2.8, the theorem is proved. $\square$

> **Corollary 4.2.10**
> *The following conditions are equivalent for a right R-module E_R:*
>
> (i) *The module E_R is injective.*
> (ii) *Every short exact sequence that begins with E_R splits, that is, every short exact sequence of right R-modules of the form $0 \to E_R \to B_R \to C_R \to 0$ splits.*
> (iii) *The module E_R is a direct summand of every module of which it is a submodule.*

Proof

(i) $\Rightarrow$ (ii) Assume E_R injective. Let $0 \to E_R \xrightarrow{f} B_R \to C_R \to 0$ be a short exact sequence. Since $f \colon E_R \to B_R$ is a monomorphism, the identity $\iota \colon E_R \to E_R$ can be factored through f, that is, there exists $g \colon B_R \to E_R$ with $g \circ f = \iota$. Hence the sequence $0 \to E_R \xrightarrow{f} B_R \to C_R \to 0$ splits.

(ii) $\Rightarrow$ (iii) Assume that (ii) holds. Let E_R be a submodule of M_R. Then the canonical exact sequence $0 \to E_R \xrightarrow{\varepsilon} M_R \to M_R/E_R \to 0$ splits by (ii), i.e., there is a morphism $f \colon M_R \to E_R$ with $f\varepsilon = \iota_M$. Then the image E_R of ε is a direct summand of M_R (Lemma 2.5.3).

(iii) $\Rightarrow$ (i) Assume that (iii) holds. By Theorem 4.2.9 there is an injective module M_R with $E_R \le M_R$. By (iii), E_R is a direct summand of the injective module M_R, so that E_R is injective by Proposition 4.2.6. □

Notice that a module M_R is: (1) projective if and only if every short exact sequence of the form $0 \to A_R \to B_R \to M_R \to 0$ splits; (2) semisimple if and only if every short exact sequence of the form $0 \to A_R \to M_R \to C_R \to 0$ splits; (3) injective if and only if every short exact sequence of the form $0 \to M_R \to B_R \to C_R \to 0$ splits.

4.3 Projective Covers

Every module is a homomorphic image of a projective module, because every module M_R is a homomorphic image of the free module $R^{(M_R)}$. Now we look for the smallest possible representation of M_R as a homomorphic image of a projective module.

4.3 Projective Covers

Definition 4.3.1 (Projective Cover)

A projective cover of a module M_R is a pair (P_R, p) where P_R is a projective right R-module and $p\colon P \to M$ is a superfluous epimorphism (that is, an epimorphism $p\colon P \to M$ with $\ker p$ a superfluous submodule of P).

Examples 4.3.2
(1) Every projective module Q has a projective cover $(Q, 1_Q)$.
(2) Let us show that a $\mathbb{Z}$-module has a projective cover if and only if it is free. If G is a free $\mathbb{Z}$-module, G has a projective cover as we have seen in the previous Example (1). Conversely, assume that the $\mathbb{Z}$-module G has a projective cover $(P_\mathbb{Z}, p)$. Then $P_\mathbb{Z}$ is free, because every projective $\mathbb{Z}$-module is free. Let $\varphi\colon P_\mathbb{Z} \to \mathbb{Z}^{(X)}$ be an isomorphism. For every $x \in X$, let $\pi_x\colon \mathbb{Z}^{(X)} \to \mathbb{Z}$ be the projection onto the x-th component, so that $\pi_x \varphi \colon P_\mathbb{Z} \to \mathbb{Z}$ is a homomorphism. Since $\pi_x \varphi(\operatorname{rad}(P_\mathbb{Z})) \subseteq \operatorname{rad}(\mathbb{Z}_\mathbb{Z}) = 0$ for every $x \in X$, we get that $\operatorname{rad}(P_\mathbb{Z}) = 0$, so that the only superfluous submodule of $P_\mathbb{Z}$ is 0. Thus $\ker(p) = 0$, i.e., p is an isomorphism. Hence G is free.
(3) Let R be a ring, P_R be a finitely generated projective R-module, N be a submodule of P_R contained in $PJ(R)$ and $\pi\colon P \to P/N$ the canonical projection. Then (P, π) is a projective cover of the right R-module $P/PJ(R)$. To prove it, it suffices to remark that $PJ(R)$ is superfluous in P_R by Nakayama's Lemma, so that the kernel N of π is superfluous in P_R because it is contained in $PJ(R)$.
(4) In particular, if R is a ring and $\pi\colon R \to R/J(R)$ is the canonical projection, then (R_R, π) is a projective cover of the right R-module $R/J(R)$.
(5) If R is a local ring, I is a proper right ideal and $\pi\colon R \to R/I$ is the canonical projection, then (R_R, π) is a projective cover of the right R-module R/I. This is also a particular case of (3).

Theorem 4.3.3
(1) (Fundamental lemma of projective covers) *Let (P, p) be a projective cover of a right R-module M. If Q is a projective module and $q\colon Q \to M$ is an epimorphism, then Q has a direct-sum decomposition $Q = P' \oplus P''$ where $P' \cong P$, $P'' \subseteq \ker(q)$ and $(P', q|_{P'}\colon P' \to M)$ is a projective cover.*
(2) (Uniqueness of projective covers up to isomorphism) *Projective covers, when they exist, are unique up to isomorphism in the following sense. If $(P, p), (Q, q)$ are any two projective covers of a right R-module M, there is an isomorphism $h\colon Q \to P$ such that $p \circ h = q$.*

Proof Let (P, p) and $q\colon Q \to M$ be as in the hypothesis of (1). As Q is projective and p is onto, there exists a morphism $h\colon Q \to P$ such that $ph = q$. Now $ph(Q) = q(Q) = M$ implies $h(Q) + \ker p = P$. (To prove this, fix $x \in P$. Then $p(x) = q(y)$ for some $y \in Q$, so that $p(x) = q(y) = ph(y)$, from which $x - h(y) \in \ker p$, hence $x \in h(Q) + \ker p$.) This implies that $h(Q) = P$, because $\ker p$ is superfluous in P. Thus h is an onto mapping. Since every homomorphism onto a projective module has a right inverse, there exists $i\colon P \to Q$ with $h \circ i = 1_P$, so that $Q = i(P) \oplus \ker(h)$. Set $P' := i(P)$ and $P'' := \ker(h)$. Then $Q = P' \oplus P''$. As $h \circ i = 1_P$ is a monomorphism, we get that i is a monomorphism, whence $P \cong i(P) = P'$. From $q = p \circ h$, it follows that $P'' = \ker(h) \subseteq \ker(q)$. In particular, $q|_{P'}(P') = q(P') = q(P' + P'') = q(Q) = M$, so that $q|_{P'}: P' \to M$ is onto. It remains to show that $\ker(q|_{P'})$ is superfluous in P'. Let us prove that $\ker(q|_{P'}) \subseteq i(\ker(p))$. If $x \in \ker(q|_{P'})$, then $x \in \ker(q)$ and $x \in P' = i(P)$. Thus $x = i(y)$ for some $y \in P$ and $q(x) = qi(y) = 0$. Now $q = p \circ h$ and $h \circ i = 1_P$ imply $qi = p$. Thus $p(y) = 0$, i.e., $y \in \ker(p)$ and $x = i(y) \in i(\ker(p))$. Thus $\ker(q|_{P'}) \subseteq i(\ker(p))$. Since the image of a superfluous submodule via a homomorphism is a superfluous submodule, we get that $i(\ker(p))$ is superfluous in Q, so that $\ker(q|_{P'})$ is superfluous in $Q = P' \oplus P''$. Thus $\ker(q|_{P'})$ is superfluous in P'. This proves (1). Now, with the same notation, suppose that (Q, q) also is a projective cover of M. As $P'' \subseteq \ker(q)$, we know that P'' is superfluous in Q. Thus $Q = P' + P''$ implies $P' = Q$, so that $P'' = 0$, i.e., $\ker(h) = 0$. Hence the surjective homomorphism h is injective as well, i.e., it is an isomorphism. □

A ring R is *right perfect* if every right R-module has a projective cover. We conclude this section with a number of characterizations of right perfect rings that appear in a famous result due to H. Bass and called *Theorem P* (see [1, Theorem 28.4]).

Theorem 4.3.4 (Bass's Theorem P)
The following conditions are equivalent for a ring R:

1. *R is right perfect.*
2. *The ring $R/J(R)$ is semisimple artinian. and the Jacobson radical $J(R)$ of R is right T-nilpotent (that is, for every sequence $x_0, x_1, x_2, \ldots$ of elements of $J(R)$ there exists an index n such that $x_n x_{n-1} \ldots x_1 x_0 = 0$).*
3. *Every non-zero right R-module has a maximal submodule and the ring $R/J(R)$ is semisimple artinian.*
4. *R satisfies the descending chain condition on principal left ideals.*
5. *Every non-zero left R-module has a simple submodule and the ring R contains no infinite orthogonal set of idempotents.*

4.4 Injective Envelopes

A submodule N of a module M_R is *essential* (or *large*) in M_R if, for every submodule L of M_R, $N \cap L = 0$ implies $L = 0$. In this case, we will write $N \leq_e M_R$.

4.4 Injective Envelopes

Exercise 4.4.1
Show that
(a) If $K \leq N \leq M_R$, then $K \leq_e M$ if and only if $K \leq_e N$ and $N \leq_e M$.
(b) If $N, N' \leq M_R$, then $N \cap N' \leq_e M$ if and only if $N \leq_e M$ and $N' \leq_e M$.
(c) The submodule M is always essential in M_R, also when $M_R = 0$.
(d) If $f : M \to M'$ is a morphism of R-modules and $N' \leq_e M'$, then $f^{-1}(N') \leq_e M$.
(e) A submodule N of an R-module M is essential in M if and only if for every $x \in M$, $x \neq 0$, there exists $r \in R$ with $xr \in N$ and $xr \neq 0$.
(f) Assume $N_1 \leq M_1 \leq M$, $N_2 \leq M_2 \leq M$ and $M = M_1 \oplus M_2$. Show that $N_1 \oplus N_2 \leq_e M_1 \oplus M_2$ if and only if $N_1 \leq_e M_1$ and $N_2 \leq_e M_2$.

A monomorphism $f : N_R \to M_R$ is said to be *essential* if its image $f(N_R)$ is an essential submodule of M_R.

Exercise 4.4.2
(a) Show that a monomorphism $f : N \to M$ is essential if and only if for every module L and every homomorphism $g : M \to L$, if gf is injective, then g is injective.
(b) Let $f : N \to M$ and $g : M \to P$ be two monomorphisms. Show that the composite mapping gf is an essential monomorphisms if and only if both f and g are essential monomorphisms.

Let M_R be a right R-module. An *extension* of M_R is a pair (N_R, f), where N_R is a right R-module and $f : M_R \to N_R$ is a monomorphism. An *essential extension* of M_R is an extension (N_R, f) where $f : M_R \to N_R$ is an essential monomorphism. An extension (N_R, f) of M_R is *proper* if f is not an isomorphism.

Proposition 4.4.3
A module M_R is injective if and only if it does not have proper essential extensions.

Proof
($\Rightarrow$) Let M_R be an injective module and $f : M_R \to N_R$ an essential monomorphism. By Corollary 4.2.10, the image of f is a direct summand of N_R, so that $N_R = f(M_R) \oplus N'_R$. Now the image of f is essential in N_R and $f(M_R) \cap N'_R = 0$, so that $N'_R = 0$. Therefore the monomorphism f is an onto mapping, that is, an isomorphism. Thus the essential extension (N_R, f) is not proper.

($\Leftarrow$) Assume that the right R-module M has no proper essential extensions. By Theorem 4.2.9, there exist an injective module Q and a monomorphism $\varepsilon : M \to Q$. Set

$$\mathcal{F} := \{ N \mid N \leq Q, \ N \cap \varepsilon(M) = 0 \}$$

It is easily seen that it is possible to apply Zorn's Lemma and find a maximal element $\overline{N}$ in $\mathcal{F}$. Let $\pi\colon Q \to Q/\overline{N}$ be the canonical projection. Consider $\pi \circ \varepsilon \colon M \to Q/\overline{N}$. We claim that $(\pi \circ \varepsilon, Q/\overline{N})$ is an essential extension of M. If we prove the claim, then $\pi \circ \varepsilon$ must be an epimorphism by hypothesis, so that $(\varepsilon(M) + \overline{N})/\overline{N} = Q/\overline{N}$. Thus $\varepsilon(M) + \overline{N} = Q$, and this sum is direct because $\overline{N} \in \mathcal{F}$. Since Q is injective, so is $\varepsilon(M) \cong M$.

It remains to prove the claim, i.e., that $(\pi \circ \varepsilon, Q/\overline{N})$ is an essential extension of M. Now $\pi \circ \varepsilon$ is a monomorphism, because $(\pi \circ \varepsilon)(x) = 0$ implies $\varepsilon(x) \in \overline{N} \cap \varepsilon(M) = 0$, so that $x = 0$ because ε is a monomorphism. Let us prove that the image of $\pi \circ \varepsilon$ is essential in $Q/\overline{N}$, i.e., that $(\varepsilon(M) \oplus \overline{N})/\overline{N} \leq_e Q/\overline{N}$. If $0 \neq X/\overline{N} \leq Q/\overline{N}$, then $X \cap \varepsilon(M) \neq 0$ by the maximality of $\overline{N}$ in $\mathcal{F}$, so that $X \cap (\varepsilon(M) \oplus \overline{N})$ properly contains $\overline{N}$. Thus $X/\overline{N} \cap (\varepsilon(M) \oplus \overline{N})/\overline{N} \neq 0$, which is what we wanted to prove.

□

Definition 4.4.4

An *injective envelope* of a module M_R is a pair (E_R, i), where E_R is an injective right R-module and $i\colon M_R \to E_R$ is an essential monomorphism. Equivalently, (E_R, i) is an essential extension of M_R with E_R an injective module.

For example, if i is the inclusion of $\mathbb{Z}_\mathbb{Z}$ into $\mathbb{Q}_\mathbb{Z}$, then $(\mathbb{Q}_\mathbb{Z}, i)$ is an injective envelope of $\mathbb{Z}_\mathbb{Z}$. Dualizing the proof of the Fundamental lemma of projective covers, we get the following

Theorem 4.4.5 (Fundamental Lemma of Injective Envelopes)

Let (E, i) be an injective envelope of a right R-module M. If F is an injective module and $j\colon M \to F$ is a monomorphism, then F has a direct-sum decomposition $F = F' \oplus F''$ where $F' \cong E$, $j(M) \subseteq F'$ and if $j'\colon M_R \to F'$ is the mapping obtained from j restricting the codomain to F', then (F', j') is an injective envelope of M.

Theorem 4.4.6

Every right R-module has an injective envelope, which is unique up to isomorphism in the following sense: if (E, i) and (E', i') are both injective envelopes of M, then there exists an isomorphism $h\colon E \to E'$ such that $hi = i'$.

4.4 Injective Envelopes

Proof

Existence. Let M be a right R-module. By Theorem 4.2.9, there exists an injective module Q containing M. Let

$$\mathcal{F} := \{ N \mid M \le N \le Q \text{ and } M \le_e N \}$$

It is easily seen that we can apply Zorn's lemma and find $\overline{N} \in \mathcal{F}$ maximal. Then $M \le_e \overline{N}$. We want to prove that $\overline{N}$ is injective, and for this it suffices to show that $\overline{N}$ has no proper essential extensions. Let (L, f) be an essential extension of $\overline{N}$. Then $f \colon \overline{N} \to L$ is an essential monomorphism. Let $\varepsilon \colon \overline{N} \hookrightarrow Q$ be the inclusion. Since Q is injective, the inclusion ε factors through f, that is, there exists $\sigma \colon L \to Q$ with $\sigma \circ f = \varepsilon$. Since $\sigma \circ f$ is injective and f is essential, σ is necessarily injective (Exercise 4.4.2(a)). The isomorphism $\sigma \colon L \to \sigma(L)$ sends the essential submodule $f(\overline{N})$ of L to the essential submodule $\sigma f(\overline{N})$ of $\sigma(L)$, that is, to the essential submodule $\varepsilon(\overline{N}) = \overline{N}$ of $\sigma(L)$. Thus $\overline{N} \le_e \sigma(L) \le Q$. By transitivity (Exercise 4.4.1(a)), $M \le_e \sigma(L)$, so that $\sigma(L) \in \mathcal{F}$. By the maximality of $\overline{N}$, we get that $\overline{N} = \sigma(L)$. Thus the essential monomorphism f is also an epimorphism, because if $l \in L$, then $\sigma(l) \in \sigma(L) = \overline{N}$, so that $f(\sigma(l)) = \varepsilon(l) = l$. Hence f is an isomorphism.

Uniqueness. Now let (E, i) and (E', i') be two injective envelopes of M. As E is injective, the homomorphism $i' \colon M \to E'$ factors through $i \colon M \to E$, that is, there exists an R-module morphism $h \colon E \to E'$ with $h \circ i = i'$. Since i is essential and $h \circ i$ is injective, h also is injective. Hence the module $h(E) \cong E$ is injective, hence $h(E)$ is a direct summand of E' (Corollary 4.2.10(iii)). Thus $E' = h(E) \oplus C$ for a suitable $C \le E'$. From $i'(M) \subseteq h(E)$, it follows that $i'(M) \cap C = 0$. But $i'(M) \le_e E'$, so that $C = 0$. Hence h is an isomorphism. □

Like for the tensor product of two modules, the injective envelope of a module M_R, which is unique up to isomorphism, will not be usually denoted as a pair. We will usually omit to indicate the embedding of M_R into the injective module. The injective envelope of a module M_R will be usually denoted by $E(M_R)$.

Examples 4.4.7
(1) Let G be a proper non-zero submodule of the Prüfer group $\mathbb{Z}(p^\infty)$, where p is a fixed prime. We know that $G \cong \mathbb{Z}/p^n\mathbb{Z}$ for a suitable $n \ge 1$. Moreover, $\mathbb{Z}(p^\infty)$ is divisible (i.e., injective) and G is essential in $\mathbb{Z}(p^\infty)$. Thus $E_\mathbb{Z}(\mathbb{Z}/p^n\mathbb{Z}) \cong \mathbb{Z}(p^\infty)$.
(2) Let R be a commutative integral domain and Q its field of fractions. Then the R-module Q_R is the injective envelope of R_R. Obviously, $R_R \le Q_R$. If $0 \ne q \in Q$ then $q = a/b$ with a and b both non-zero, so $bq = a \in R$ is non-zero. Thus R is essential in Q.
We must now show that Q is injective, and for this we will apply Baer's criterion. Let $I \trianglelefteq R$, and let $\sigma \colon I \to Q$ be an R-module morphism. We must prove that it can be extended to R. If $I = 0$, then $\sigma = 0$, so that the zero mapping $R \to Q$ extends σ. Thus we can assume $I \ne 0$. Take any two non-zero elements $a, b \in I$. From $a\sigma(b) = \sigma(ab) = b\sigma(a)$, it follows that $\sigma(b)/b = \sigma(a)/a$.

That is, $\sigma(a)/a$ doesn't depend on the non-zero element a of I, it depends only on σ. Let $q_\sigma \in Q$ be this constant, so that $\sigma(a) = qa$ for every $a \in I$, $a \neq 0$. Then trivially $\sigma(a) = qa$ also for $a = 0$, so that multiplication by q is an R-module morphism $\lambda_q \colon R \to Q$ that extends σ.

The next lemma follows immediately from Exercise 4.4.2(b).

Lemma 4.4.8
If R is a ring and $N_R \leq_e M_R$, then $E(N_R) = E(M_R)$. More precisely, if (M, f) is an essential extension of N and (E, ε) is an injective envelope of M, then $(E, \varepsilon \circ f)$ is an injective envelope of N.

Proposition 4.4.9
An extension (E, ε) of a module M is an injective envelope of M if and only if it is a maximal essential extension of M. More precisely, let $\varepsilon \colon M \to E$ be a right R-module monomorphism. Then (E, ε) is an injective envelope of M if and only if it is an essential extension and, for every monomorphism $f \colon E \to N$, if $(N, f \circ \varepsilon)$ is an essential extension of M, then f is an isomorphism.

Proof

($\Rightarrow$) Let (E, ε) be an injective envelope of M, i.e., an essential extension of M with E injective. Let $f \colon E \to N$ be a monomorphism with $(N, f \circ \varepsilon)$ an essential extension of M. From Exercise 4.4.2(b), we get that (N, f) is an essential extension of E, which cannot be proper because E is injective (Proposition 4.4.3). Hence f is an isomorphism.

($\Leftarrow$) Let (E, ε) be a maximal essential extension of M. It suffices to show that E is injective, i.e., that it does not have proper essential extensions. Fix an essential extension (N, f) of E. Then f is a monomorphism and $(N, f \circ \varepsilon)$ is an essential extension of M (Exercise 4.4.2(b)). By hypothesis, f is an isomorphism. □

Thus we have that the injective envelope (E, ε) of a right R-module M can be defined in three different equivalent ways, as follows.

4.4 Injective Envelopes

Theorem 4.4.10
Let M be a right R-module. Then there exists an extension (E, ε) of M satisfying the following equivalent conditions:

(a) (E, ε) is an essential injective extension of M.
(b) (E, ε) is a maximal essential extension of M.
(c) (E, ε) is a minimal injective extension of M.
Moreover, such an extension is unique up to isomorphism in the following sense. If both (E, ε) and (E', ε') are extensions of M satisfying these three equivalent conditions, then there exists an isomorphism $\varphi \colon E \to E'$ such that $\varphi \varepsilon = \varepsilon'$.

From Exercise 4.4.1(f) and Proposition 4.2.7 we immediately get that:

Proposition 4.4.11
If $M_1, \ldots, M_n$ are modules and $M = M_1 \oplus \cdots \oplus M_n$, then $E(M) = E(M_1) \oplus \cdots \oplus E(M_n)$.

Proposition 4.4.12
For every submodule N of an R-module M there exists a submodule C of M with $C \cap N = 0$ and $N \oplus C \leq_e M$.

Proof Let N be a submodule of M. Set

$$\mathcal{F} := \{ X \leq M \mid X \cap N = 0 \}.$$

Then $\mathcal{F}$ is non-empty because it contains the zero submodule. Partial order $\mathcal{F}$ by inclusion. We can apply Zorn's lemma, so that there exists an element C maximal in $\mathcal{F}$. It suffices to show that $N \oplus C$ is essential in M. Let $X \leq M$, and assume that $(N \oplus C) \cap X = 0$. Then we have an internal direct sum $N \oplus C \oplus X$, which is a submodule of M. Thus $C \oplus X \in \mathcal{F}$. By the maximality of C, it follows that $X = 0$. □

Proposition 4.4.13
If $N \leq M$, then $E(N)$ is a direct summand of $E(M)$.

Proof Let $N \leq M$. By Proposition 4.4.12 there is a submodule $C \leq M$ with $C \cap N = 0$ and $N \oplus C$ essential in M. By Lemma 4.4.8 and Proposition 4.4.11,

$$E(M) = E(N \oplus C) \cong E(N) \oplus E(C).$$

Hence $E(N)$ is a direct summand of $E(M)$. □

From Lemma 3.4.5, we know that the radical of a module, i.e., the intesection of all maximal submodules, is the sum of all superfluous submodules. In the next proposition we prove the dual result. Recall that the socle is the sum of all simple submodules, that is, the sum of all minimal submodules.

Proposition 4.4.14
For every module M, soc(M) is the intersection of all essential submodules of M.

Proof Set $I := \bigcap_{N \leq_e M} N$. In order to prove that soc(M) $\subseteq I$, it suffices to show that if S is a simple submodule of M and $N \leq_e M$, then $S \leq N$. Now if S is a simple submodule of M and $N \leq_e M$, then $S \cap N$ is a submodule of S, so that either $S \cap N = 0$ or $S \cap N = S$. Now if $S \cap N = 0$, then $S = 0$ because N is essential, but $S = 0$ contradicts the fact that S is simple. Hence $S \cap N = S$, that is, $S \subseteq N$.

In order to prove that $I \subseteq$ soc(M), it suffices to show that I is semisimple. Now if $N \leq I$, by Proposition 4.4.12 there exists a submodule C of M with $C \cap N = 0$ and $N \oplus C \leq_e M$. Thus $I \subseteq N \oplus C$. From Exercise 1.8.1, it follows that $I = N \oplus (C \cap I)$. Thus N is a direct summand of I, and I is semisimple. □

4.5 Uniform Modules, Goldie Dimension

Lemma 4.5.1
The following conditions are equivalent for a non-zero module M_R:
(i) If $N, N' \leq M_R$ and $N \cap N' = 0$, then $N = 0$ or $N' = 0$.
(ii) The intersection of two non-zero submodules of M_R is non-zero.
(iii) Every non-zero submodule of M_R is essential in M_R.
(iv) The injective envelope $E(M_R)$ of M_R is an indecomposable module.

Proof The equivalences (i) $\Leftrightarrow$ (ii) and (ii) $\Leftrightarrow$ (iii) are trivial.

(ii) $\Rightarrow$ (iv). Assume (iv) false, so that $E(M_R) = E_1 \oplus E_2$ with E_1 and E_2 both non-zero. Since M_R is essential in $E(M_R)$, it follows that $E_1 \cap M_R$ and $E_2 \cap M_R$ are both non-zero. But $(E_1 \cap M_R) \cap (E_2 \cap M_R) \subseteq E_1 \cap E_2 = 0$. This contradicts (ii).

(iv) $\Rightarrow$ (i). Suppose $E(M_R)$ indecomposable, $N, N' \leq M_R$ and $N \cap N' = 0$. Then $N \oplus N'$ is a submodule of M_R, so that $E(M_R) = E(N \oplus N') \oplus C$ by Proposition 4.4.13. Thus $E(M_R) = E(N) \oplus E(N') \oplus C$ (Proposition 4.4.11). Since $E(M_R)$ is indecomposable, two of the three modules $E(N), E(N'), C$ are zero. Hence at least one of the two modules $E(N), E(N')$ is zero. A fortiori, one of the two modules N, N' is zero.

□

A module M_R is *uniform* if it is non-zero and satisfies the equivalent conditions of Lemma 4.5.1. From condition (i) one sees immediately that every uniform module is indecomposable.

Proposition 4.5.2
The following conditions are equivalent for an injective module E_R:
(i) *E_R is indecomposable.*
(ii) *E_R is uniform.*
(iii) *The endomorphism ring of E_R is local.*

Proof
(i) $\Rightarrow$ (ii) Let E_R be an indecomposable injective module. Then E_R coincides with its injective envelope $E(E_R)$, so that E_R satisfies condition (iv) of Lemma 4.5.1. Hence E_R is a uniform module.

(ii) $\Rightarrow$ (iii) Let E_R be a uniform injective module. We claim that an endomorphism φ of E_R is an automorphism of E_R if and only if it is a monomorphism. If $\varphi \colon E_R \to E_R$ is a monomorphism, then $\varphi(E_R)$ is isomorphic to E_R, hence it is injective and non-zero. Thus $\varphi(E_R)$ is an injective submodule of E_R, hence $\varphi(E_R)$ is a direct summand of E_R, that is, $E_R = \varphi(E_R) \oplus C$ for some submodule C of E_R. As E_R is uniform, it is indecomposable, and $\varphi(E_R) \neq 0$. Thus $C = 0$, and $E_R = \varphi(E_R)$. This shows that φ is also an epimorphism, hence it is an automorphism of E_R. This proves our claim. In order to show that the ring $\text{End}(E_R)$ is local, it suffices to show that the sum of two non-invertible elements of $\text{End}(E_R)$ is non-invertible, that is, that the sum of two endomorphisms of E_R that are not automorphisms of E_R is not an automorphism of E_R. Now if $\varphi, \psi \colon E_R \to E_R$ are two endomorphisms of E_R that are not automorphisms, then, by the claim, φ and ψ are not monomorphisms, that is $\ker(\varphi) \neq 0$ and $\ker(\psi) \neq 0$. Now E_R is uniform, so that, by Lemma 4.5.1(ii), $0 \neq \ker(\varphi) \cap \ker(\psi)$. But $\ker(\varphi) \cap \ker(\psi) \subseteq \ker(\varphi + \psi)$, so that $\varphi + \psi$ is not a monomorphism, as desired.

(iii) $\Rightarrow$ (i)   follows from Proposition 4.1.7. $\square$

Let M_R be a right module. A set $\{ N_\lambda \mid \lambda \in \Lambda \}$ of non-zero submodules of M_R is said to be *independent* if its sum is direct:

$$\sum_{\lambda \in \Lambda} N_\lambda = \bigoplus_{\lambda \in \Lambda} N_\lambda.$$

The empty set is an independent set of non-zero submodules of M_R. By Zorn's Lemma every independent set of non-zero submodules of M_R is contained in a maximal independent set of non-zero submodules of M_R.

Lemma 4.5.3
Let M_R be a module without uniform submodules. Then M_R has an infinite independent set of non-zero submodules.

Proof Let M_R be a module without uniform submodules. We will define a sequence $A_1, A_2, A_3, \ldots$ of non-zero submodules of M_R such that, for every $n \geq 1$, the set $\{A_1, A_2, \ldots, A_n\}$ is independent and $A_1 \oplus \cdots \oplus A_n$ is not essential in M_R. The construction of the submodules A_n is by induction on n. For $n = 1$, note that M_R is not uniform, hence there exist non-zero submodules A_1, A_1' of M_R such that $A_1 \cap A_1' = 0$, i.e., A_1 has the required properties. Suppose $A_1, \ldots, A_{n-1}$ have been constructed. Since $A_1 \oplus \cdots \oplus A_{n-1}$ is not essential in M_R, there exists a non-zero $B \leq M_R$ such that $B \cap (A_1 \oplus \cdots \oplus A_{n-1}) = 0$. The module B is not uniform. Hence there exist $A_n, A_n' \leq B$, where A_n and A_n' are non-zero, such that $A_n \cap A_n' = 0$. Then $A_n \cap (A_1 \oplus \cdots \oplus A_{n-1}) = 0$, so that $\{A_1, A_2, \ldots, A_n\}$ is independent. Moreover $A_n' \cap (A_1 \oplus \cdots \oplus A_n) = 0$, because if $a_n' = a_1 + \cdots + a_n$, then $a_n' - a_n = a_1 + \cdots + a_{n-1} \in B \cap (A_1 \oplus \cdots \oplus A_{n-1}) = 0$. Thus $a_n = a_n' \in A_n \cap A_n' = 0$. This completes the construction of the modules A_n. Now $\{ A_n \mid n \geq 1 \}$ is an infinite independent set of non-zero submodules of M_R. $\square$

4.5 Uniform Modules, Goldie Dimension

Theorem 4.5.4
The following conditions are equivalent for a module M_R.

(a) *M_R does not have an infinite independent set of non-zero submodules.*
(b) *M_R has a finite independent set $\{A_1, A_2, \ldots, A_n\}$ of uniform submodules and $A_1 \oplus \cdots \oplus A_n$ is essential in M_R.*
(c) *There exists a non-negative integer m such that the cardinalities of all the independent sets of non-zero submodules of M_R are $\leq m$.*
(d) *If $A_0 \leq A_1 \leq A_2 \leq \ldots$ is an ascending chain of submodules of M_R, then there exists $i \geq 0$ such that A_i is essential in A_j for every $j \geq i$.*

Moreover, if these equivalent conditions hold and $\{A_1, A_2, \ldots, A_n\}$ is a finite independent set of uniform submodules with $A_1 \oplus \cdots \oplus A_n$ essential in M_R, then any other independent set of non-zero submodules of M_R has cardinality $\leq n$.

Proof

(a) $\Rightarrow$ (b). Let $\mathcal{F}$ be the family of all independent subsets of uniform submodules of M_R. The family $\mathcal{F}$ is non-empty (Lemma 4.5.3). By Zorn's Lemma, the family $\mathcal{F}$ has a maximal element X with respect to inclusion. By (a), the set X is finite, $X = \{A_1, A_2, \ldots, A_n\}$ say. The element $A_1 \oplus \cdots \oplus A_n$ is essential in M_R, otherwise there would exist a non-zero submodule B of M_R such that $(A_1 \oplus \cdots \oplus A_n) \cap B = 0$, and so, by Lemma 4.5.3, there would be a uniform submodule C of M_R such that $C \leq B$. Then $\{A_1, A_2, \ldots, A_n, C\}$ would be independent, a contradiction.

(b) $\Rightarrow$ (c). Suppose that (b) holds, so that there exists a finite independent set $\{A_1, A_2, \ldots, A_n\}$ of uniform submodules of M_R with $A_1 \oplus \cdots \oplus A_n$ essential in M_R. Assume that there exists an independent set $\{B_1, B_2, \ldots, B_k\}$ of non-zero submodules of M_R of cardinality $k > n$. We will reach a contradiction. For every $t = 0, 1, \ldots, n$, we will construct a subset X_t of $\{A_1, A_2, \ldots, A_n\}$ of cardinality t and a subset Y_t of $\{B_1, B_2, \ldots, B_k\}$ of cardinality $k - t$ such that $X_t \cap Y_t = \emptyset$ and $X_t \cup Y_t$ is independent. For $t = 0$, set $X_0 := \emptyset$ and $Y_0 := \{B_1, B_2, \ldots, B_k\}$. Suppose that X_t and Y_t have been constructed for some $t, 0 \leq t < n$. We will construct X_{t+1} and Y_{t+1}. Since $|Y_t| = k - t > n - t > 0$, there exists $j \in \{1, 2, \ldots, k\}$ with $B_j \in Y_t$. But $X_t \cup Y_t$ is independent, so that we have a direct sum

$$C := (\oplus_{A \in X_t} A) \oplus (\oplus_{B \in Y_t \setminus \{B_j\}} B).$$

We claim that $C \cap A_\ell = 0$ for some $\ell \in \{1, 2, \ldots, n\} \setminus X_t$. Otherwise, if $C \cap A_\ell \neq 0$ for every $\ell \in \{1, 2, \ldots, n\} \setminus X_t$, then $C \cap A_\ell \neq 0$ for every $\ell = 1, \ldots, n$, so that $C \cap A_\ell$ is essential in A_ℓ because A_ℓ is uniform. Thus $\oplus_{\ell=1}^n C \cap A_\ell$ is essential in $\oplus_{\ell=1}^n A_\ell$ by Exercise 4.4.1(f). As $\oplus_{\ell=1}^n A_\ell$ is essential in M_R, it follows that $\oplus_{\ell=1}^n C \cap A_\ell$ is essential in M_R (Exercise 4.4.1(a)). Then $C \supseteq \oplus_{\ell=1}^n C \cap A_\ell$ is essential in M_R, so that $C \cap B_j \neq 0$. This contradicts the fact that $X_t \cup Y_t$ is independent and this contradiction proves the claim. From the claim, it follows that $(X_t \cup \{A_\ell\}) \cup (Y_t \setminus \{B_j\})$ is independent, so that $X_{t+1} := X_t \cup \{A_\ell\}$ and $Y_{t+1} := Y_t \setminus \{B_j\}$ have the required properties. This completes the construction of the sets X_t and Y_t.

For $t = n$, we have a non-empty subset Y_n of $\{B_1, B_2, \ldots, B_k\}$ such that $\{A_1, A_2, \ldots, A_n\} \cup Y_n$ is an independent set of cardinality k of submodules, so that

$$(A_1 \oplus A_2 \oplus \cdots \oplus A_n) \cap B = 0$$

for every $B \in Y_n$. This contradicts the fact that $A_1 \oplus A_2 \oplus \cdots \oplus A_n$ is essential in L. Hence every independent set of non-zero submodules of M_R has cardinality $\leq n$.

Notice that, in this proof of (b) $\Rightarrow$ (c), we have also proved the last part of the statement of the Theorem.

(c) $\Rightarrow$ (d). If (d) does not hold, there is a chain $A_0 \leq A_1 \leq A_2 \leq \ldots$ of submodules of M_R such that for every $i \geq 0$ there exists $j(i) > i$ with A_i not essential in $A_{j(i)}$. Set $j_0 = 0$ and $j_{n+1} = j(j_n)$ for every $n \geq 0$. Then for every $n \geq 0$ there exists a non-zero submodule B_n of $A_{j_{n+1}}$ such that $B_n \cap A_{j_n} = 0$. The set $\{B_n \mid n \geq 0\}$ is obviously independent. Thus (c) does not hold.

(d) $\Rightarrow$ (a). If (a) is not satisfied, then M_R contains a countable infinite independent set $\{B_i \mid i \geq 0\}$ of non-zero submodules. Set $A_n = \oplus_{i=0}^n B_i$. Then $A_0 \leq A_1 \leq A_2 \leq \ldots$, and for every $n \geq 0$ the element A_n is not essential in A_{n+1} because

$$A_n \cap B_{n+1} = 0.$$

Hence (d) is not satisfied.

$\square$

Thus, for a module M_R, either there is a finite independent set $\{A_1, A_2, \ldots, A_n\}$ of uniform submodules of M_R with

$$A_1 \oplus A_2 \oplus \cdots \oplus A_n$$

essential in M_R, and in this case n is said to be the *Goldie dimension* $\dim M_R$ of M_R, or M_R contains infinite independent sets of non-zero submodules, in which case M_R is said to have *infinite Goldie dimension*. For instance, uniform modules are exactly the modules of Goldie dimension one.

Since a module M is essential in its injective envelope $E(M)$,

$$\dim(M) = \dim(E(M)).$$

4.5 Uniform Modules, Goldie Dimension

If a module M has finite Goldie dimension n, it contains an essential submodule that is the finite direct sum of n uniform submodules $U_1, \ldots, U_n$, and in this case $E(M) = E(U_1) \oplus E(U_2) \oplus \cdots \oplus E(U_n)$ is the finite direct sum of n indecomposable modules. Note that, by Theorem 4.1.13, we already knew that if $E(M)$ is a finite direct sum of indecomposable modules, then the number of direct summands in any indecomposable decomposition of $E(M)$ does not depend on the decomposition. Hence a module M has finite Goldie dimension n if and only if its injective envelope $E(M)$ is the direct sum of n indecomposable modules.

In the next proposition, we collect the most important arithmetical properties of the Goldie dimension of modules. Some of these properties have already been noticed. Their proof is elementary.

Proposition 4.5.5
Let M be module.

(a) $\dim(M) = 0$ if and only if $M = 0$.
(b) $\dim(M) = 1$ if and only if M is uniform.
(c) If $N \leq M$ and M has finite Goldie dimension, then N has finite Goldie dimension and $\dim(N) \leq \dim(M)$.
(d) If $N \leq M$ and M has finite Goldie dimension, then $\dim(N) = \dim(M)$ if and only if N is essential in M.
(e) If M and M' are modules of finite Goldie dimension, then $M \oplus M'$ is a module of finite Goldie dimension $\dim(M \oplus N) = \dim(M) + \dim(N)$.

Artinian modules and noetherian modules have finite Goldie dimension. For an artinian module M, the Goldie dimension of M is equal to the composition length of its socle $\text{soc}(M)$. In particular, an artinian module M has Goldie dimension 1 if and only if it has a simple socle.

We conclude with a remark. The proper setting of Goldie dimension is lattice theory, because Goldie dimension can be defined for every bounded modular lattice, as follows.

Let $(L, \vee, \wedge)$ be a lattice. Recall that the lattice $(L, \vee, \wedge)$ is *bounded* if is has a smallest element, usually denoted by 0, and a greatest element, usually denoted by 1. A lattice $(L, \vee, \wedge)$ is *modular* if it satisfies the *modular identity* $a \wedge (b \vee c) = (a \wedge b) \vee c$ for every $a, b, c \in L$ with $c \leq a$. For every module M_R, the lattice $\mathcal{L}(M)$ of all submodules of M_R is bounded and modular. For the rest of this Section, let L be a bounded modular lattice. If $a, b \in L$ and $a \leq b$ let $[a, b] = \{x \in L \mid a \leq x \leq b\}$ be the *interval* between a and b.

A finite subset $\{a_i \mid i \in I\}$ of $L \setminus \{0\}$ is said to be *join-independent* if $a_i \wedge (\bigvee_{j \neq i} a_j) = 0$ for every $i \in I$. The empty subset of $L \setminus \{0\}$ is join-independent. An arbitrary subset A of $L \setminus \{0\}$ is *join-independent* if all its finite subsets are join-independent.

By Zorn's Lemma, every join-independent subset of $L \setminus \{0\}$ is contained in a maximal join-independent subset of $L \setminus \{0\}$.

An element $a \in L$ is *essential* in L if $a \wedge x \neq 0$ for every non-zero element $x \in L$. Thus 0 is essential in L if and only if $L = \{0\}$. If a, b are elements of L, $a \leq b$ and a is essential in the lattice $[0, b]$, then a is said to be *essential* in b. In particular, 0 is essential in b if and only if $b = 0$.

A lattice $L \neq \{0\}$ is *uniform* if all its non-zero elements are essential in L, that is, if $x, y \in L$ and $x \wedge y = 0$ imply $x = 0$ or $y = 0$. An element a of a modular lattice L is *uniform* if $a \neq 0$ and the lattice $[0, a]$ is uniform.

Theorem 4.5.6
The following conditions are equivalent for a modular lattice L with 0.

(a) *L does not contain infinite join-independent subsets.*
(b) *L contains a finite join-independent subset $\{a_1, a_2, \ldots, a_n\}$ with a_i uniform for every $i = 1, 2, \ldots, n$ and $a_1 \vee a_2 \vee \cdots \vee a_n$ essential in L.*
(c) *The cardinality of the join-independent subsets of L is $\leq m$ for a non-negative integer m.*
(d) *If $a_0 \leq a_1 \leq a_2 \leq \ldots$ is an ascending chain of elements of L, then there exists $i \geq 0$ such that a_i is essential in a_j for every $j \geq i$.*

Moreover, if these equivalent conditions hold and $\{a_1, a_2, \ldots, a_n\}$ is a finite join-independent subset of L with a_i uniform for every $i = 1, 2, \ldots, n$ and $a_1 \vee a_2 \vee \cdots \vee a_n$ essential in L, then any other join-independent subset of L has cardinality $\leq n$.

Thus, for a modular lattice L, either there is a finite join-independent subset $\{a_1, a_2, \ldots, a_n\}$ with a_i uniform for every $i = 1, 2, \ldots, n$ and

$$a_1 \vee a_2 \vee \cdots \vee a_n$$

essential in L, and in this case n is said to be the *Goldie dimension* $\dim L$ of L, or L contains infinite join-independent subsets, in which case L is said to have *infinite Goldie dimension*. The Goldie dimension of a lattice L is zero if and only if L has exactly one element.

4.6 Direct Limits of Modules

Let R be a ring, $(I, \leq)$ a preordered set and $\{M_i \mid i \in I\}$ be a family of right R-modules indexed in I. For every $i, j \in I$ with $i \leq j$, let $\mu_{ij} \colon M_i \to M_j$ be a module morphism. Suppose that:

(1) for every $i, j, k \in I$ with $i \leq j \leq k$, $\mu_{jk} \circ \mu_{ij} = \mu_{ik}$, and
(2) $\mu_{ii} = \iota_{M_i}$ for every $i \in I$.

Then the modules $(M_i)_{i \in I}$ together with the morphisms $(\mu_{ij})_{i \leq j}$ are said to form a *direct system* $\mathbb{M} := (M_i, \mu_{ij})_{i,j}$ of right R-modules over the preordered set I.

▶ **Remark 4.6.1** A direct system of right R-modules is essentially a functor $I \to \text{Mod}-R$, where I is the category associated to the preordered set $(I, \leq)$ (Example 1.2.2(7)). More precisely, we can construct a category whose objects are all direct systems of right R-modules over a preordered set I (we will construct this category on page 161) and a category whose objects are all functors $I \to \text{Mod}-R$ (the morphisms between two functors $I \to \text{Mod}-R$ are the natural transformations between them.) Then the two categories turn out to be isomorphic.

Given any direct system $\mathbb{M} = (M_i, \mu_{ij})_{i,j}$ over the preordered set I, we will construct an R-module M, called the *direct limit* (or *inductive limit*, or *colimit*) of the system. More precisely, the direct limit will be a pair (M, μ_i), where M is a module and the morphisms μ_i, one for each $i \in I$, form a family of morphisms $\mu_i \colon M_i \to M$.

Set $C := \oplus_{i \in I} M_i$. We will identify each module M_i with its image in C. Let D be the submodule of C generated by all elements of the form $x_i - \mu_{ij}(x_i)$, where $i, j \in I$, $i \leq j$ and $x_i \in M_i$ for every $i \in I$. Define the direct limit to be the module $M := C/D$ with the morphisms $\mu_i \colon M_i \to M$, defined by $\mu_i(x_i) = x_i + D$. Equivalently, the morphisms μ_i are the restrictions of the canonical projections $C \to C/D = M$ to the submodules M_i's. The direct limit M will be denoted by

$$\varinjlim M_i.$$

Notice that, if $i, j \in I$ and $i \leq j$, then $\mu_j \circ \mu_{ij} = \mu_i$, because $x_i - \mu_{ij}(x_i) \in D$ for every $i \leq j$.

A preordered set[1] $(I, \leq)$ is said to be a *directed set* if, for every $i, j \in I$, there exists $k \in I$ such that $i \leq k$ and $j \leq k$.

---

[1] We have seen in Exercise 1.2.4 the relation between preorders and partial orders. As a consequence (Exercise 4.6.8), there is no serious difference in studying direct systems or direct limits of modules over preordered sets or partially ordered sets. Hence the reader can suppose that I is always a partially ordered set, if this helps him to feel more at ease.

In what we have seen until now, the fact that the preordered set I is directed was not necessary (direct systems and direct limits can be defined and constructed over any preordered set I, also when I is not directed). In the next Lemma, we need I to be a directed set.

Lemma 4.6.2
Let I be a directed set. Then, in the notation above, every element of the direct limit M can be written as $\mu_i(x_i)$ for suitable elements $i \in I$ and $x_i \in M_i$. That is, M is the union of the images of the morphisms μ_i, where i ranges in I.

Proof An element of M is of the form $x_{i_1} + \cdots + x_{i_n} + D$, where $n \geq 1, i_1, \ldots, i_n \in I$, and $x_{i_t} \in M_{i_t}$ for every $t = 1, \ldots, n$. Let $i \in I$ be such that $i_t \leq i$ for every $t = 1, \ldots, n$ (such an i exists because I is directed). Then $x_{i_t} - \mu_{i_t i}(x_{i_t}) \in D$ for every $t = 1, \ldots, n$. Set $x_i := \mu_{i_1 i}(x_{i_1}) + \cdots + \mu_{i_n i}(x_{i_n}) \in M_i$. Then

$$x_{i_1} + \cdots + x_{i_n} + D = \mu_{i_1 i}(x_{i_1}) + \cdots + \mu_{i_n i}(x_{i_n}) + D = x_i + D = \mu_i(x_i).$$

□

Example 4.6.3
Let I be a set of cardinality at least 2 and consider the equality relation $=$ on the set I. Then $(I, =)$ is a partially ordered set, which is not directed. Direct systems of right R-modules over I coincide essentially with families $\{ M_i \mid i \in I \}$ of right R-modules indexed in I. The direct limit of the direct system corresponding to the family $\{ M_i \mid i \in I \}$ is the direct sum $M = \oplus_{i \in I} M_i$ of the family, and the morphisms $\mu_i : M_i \to M$ are the canonical inclusions of the modules M_i in their direct sum. Notice that Lemma 4.6.2 does not hold in this case.

Lemma 4.6.4
In the notations of this Section, assume that I is a directed set. Let $i \in I$ and $x_i \in M_i$ be fixed elements. Then $\mu_i(x_i) = 0$ if and only if there is a $j \geq i$ such that $\mu_{ij}(x_i) = 0$.

Proof
($\Rightarrow$) If $\mu_i(x_i) = 0$, then $x_i \in D$. This means that x_i is a linear combinations of the generators of D with which we have defined D, i.e., there exist $n \geq 1$, $r_1, \ldots, r_n \in R$, $i_1, \ldots, i_n, j_1, \ldots, j_n \in I$ with $i_t \leq j_t$ and $x_{i_t} \in M_{i_t}$ for every $t = 1, \ldots, n$, such that

$$x_i = \sum_{t=1}^n (x_{i_t} - \mu_{i_t j_t}(x_{i_t})) r_t \qquad (4.1)$$

4.6 Direct Limits of Modules

in C. Let $j \in I$ be such that $i, i_1, \ldots, i_n, j_1, \ldots, j_n \leq j$. Let $\pi : C \to M_j$ be the right R-module morphism defined by

$$\pi((x_i)_{i \in I}) = \sum_{\substack{i \in I \\ i \leq j}} \mu_{ij}(x_i).$$

Applying the module morphism π to the equality (4.1), we get that

$$\mu_{ij}(x_i) = \pi(x_i) = \sum_{t=1}^{n} (\mu_{i_t j}(x_{i_t}) - \mu_{j_t j}(\mu_{i_t j_t}(x_{i_t}))) = 0$$

by the compatibility of the morphisms μ_{ij}.

($\Leftarrow$) If $\mu_{ij}(x_i) = 0$, then $\mu_i(x_i) = \mu_j(\mu_{ij}(x_i)) = 0$.

□

Direct limit is characterized by the following *Universal Property of Direct Limit*:

> **Proposition 4.6.5**
> Let I be a preordered set, $\mathbb{M} = (M_i, \mu_{ij})$ a direct system of right modules over a ring R, and (M, μ_i) its direct limit. Let N_R be a right R-module and let $\alpha_i : M_i \to N$, $i \in I$, be a family of right R-module morphisms such that $\alpha_i = \alpha_j \circ \mu_{ij}$ whenever $i \leq j$. Then there is a unique morphism $\alpha : M \to N$ such that $\alpha_i = \alpha \circ \mu_i$ for every $i \in I$.

Proof
Existence of the morphism α.   The morphism

$$\alpha' : C = \bigoplus_{i \in I} M_i \longrightarrow N$$

$$(x_i)_{i \in I} \longmapsto \sum_{i \in I} \alpha_i(x_i),$$

applied to any generator $x_i - \mu_{ij}(x_i)$ of D, yields that

$$\alpha'(x_i - \mu_{ij}(x_i)) = \alpha_i(x_i) - \alpha_j \mu_{ij}(x_i) = 0.$$

Thus $\alpha'(D) = 0$, so that α' induces a morphism $\alpha : M = C/D \to N$, defined by $\alpha(\sum_t x_{i(t)} + D) = \sum_t \alpha_{i(t)}(x_{i(t)})$.

Let us show that α has the required property $\alpha_i = \alpha \circ \mu_i$ for every $i \in I$. For every $i \in I$ and $x_i \in M_i$, we have that

$$(\alpha \circ \mu_i)(x_i) = \alpha(\mu_i(x_i)) = \alpha(x_i + D) = \alpha'(x_i) = \alpha_i(x_i).$$

Uniqueness. Let $\beta \colon M \to N$ be any morphism with $\alpha_i = \beta \circ \mu_i$ for every $i \in I$. Consider an element x of M. Such an element is necessarily of the form $x = x_{i_1} + \cdots + x_{i_n} + D$ for suitable $x_{i_j} \in M_j$. Then, for every $j = 1, \ldots, n$, we have that

$$\alpha(x_{i_j} + D) = (\alpha \circ \mu_j)(x_{i_j}) = \alpha_j(x_{i_j}) = (\beta \circ \mu_j)(x_{i_j}) = \beta(x_{i_j} + D),$$

and so $\alpha(x) = \beta(x)$.

$\square$

▶ **Remark 4.6.6** The Universal Property 4.6.5 characterizes direct limit in the following sense. Let (M, μ_i) be the direct limit. Then (M, μ_i) satisfies the Universal Property 4.6.5. If (M', μ'_i) also satisfies the Universal Property 4.6.5, that is,

(i) M' is a module,
(ii) the $\mu'_i \colon M_i \to M'$ are morphisms, $\mu'_j \mu_{ij} = \mu'_i$ for every $i \leq j$, and
(iii) for any module N and any family of morphisms $\alpha_i \colon M_i \to N$ with $\alpha_i = \alpha_j \circ \mu_{ij}$ for all $i \leq j$, there is a unique morphism $\alpha \colon M' \to N$ such that $\alpha_i = \alpha \circ \mu'_i$ for every $i \in I$,

then M' is canonically isomorphic to M. More precisely, there is a unique isomorphism $\varphi \colon M \to M'$ such that $\varphi \mu_i = \mu'_i$ for any $i \in I$. We leave the proof of this fact to the reader.

Direct limit is unique up to isomorphism because it satisfies a universal property, i.e., it is a solution of a universal problem. We will return to solutions of universal problems when we introduce initial objects in categories.

Example 4.6.7
Let M_R be a module, I be a non-empty set and M_i, $i \in I$, a family of submodules of M_R with the property that, for every $i, j \in I$, there exists $k \in I$ such that $M_i + M_j \subseteq M_k$. We can define a relation $\leq$ on the set I by setting, for any $i, j \in I$, $i \leq j$ if $M_i \subseteq M_j$. It is easily seen that $(I, \leq)$ is a directed preordered set. We can prove that $\varinjlim M_i = \bigcup_{i \in I} M_i$ making use of the uniqueness of the solution of the Universal Property (Proposition 4.6.5). Consider the embeddings $\varepsilon_i \colon M_i \to \bigcup_{k \in I} M_k$ and $\varepsilon_{ij} \colon M_i \to M_j$ for all $i \leq j$, so that $\varepsilon_j \circ \varepsilon_{ij} = \varepsilon_i$. Let N_R be any right R-module with morphisms $\nu_j \colon M_j \to N$ such that $\nu_j \circ \varepsilon_{ij} = \nu_i$ for every $i \leq j$. Define a mapping $\nu \colon \bigcup_{k \in I} M_k \to N_R$ as follows. For any $x \in \bigcup_{k \in I} M_k$, there exists an index $i \in I$ with $x \in M_i$. Set $\nu(x) := \nu_i(x)$. This map is well-defined, because if $x \in M_i$ and $x \in M_j$, there exists an index k with $i \leq k$ and $j \leq k$. Then $\nu_i(x) = \nu_k(\varepsilon_{ik}(x)) = \nu_k(x)$ and similarly $\nu_j(x) = \nu_k(x)$. Thus ν is well defined, and clearly its restriction to any M_i is ν_i, that is, $\nu \varepsilon_i = \nu_i$. Finally, any other mapping $\nu' \colon \bigcup_{k \in I} M_k \to N_R$ such that $\nu' \varepsilon_i = \nu_i$ for all $i \in I$ must coincide with ν_i on M_i, hence must be equal to ν. Thus the union $\bigcup_{i \in I} M_i$ with the morphisms ε_i is a solution of the universal problem.

4.6 Direct Limits of Modules

From the previous example, we see that any right R-module is the direct limit of a family of finitely generated submodules.

Exercise 4.6.8
Let I be a directed preordered set, $\mathbb{M} = (M_i, \mu_{ii'})$ a direct system of right R-modules over I, and (M, μ_i) its direct limit. Let $J \subseteq I$ be a *cofinal* subset of I, that is, a subset of I such that for every $i \in I$ there exists $j \in J$ with $i \leq j$.
(a) Show that J is directed.
(b) Let $\mathbb{N} = (N_j, \nu_{jj'})$ be the direct system with $N_j = M_j$ for every $j \in J$ and $\nu_{jj'} = \mu_{jj'}$ for every $j, j' \in J$, $j \leq j'$. Let (N, ν_i) be its direct limit. Show that there is an isomorphism $\varphi \colon M \to N$ such that $\varphi \mu_j = \nu_j$ for every $j \in J$.

Notice that any preordered set contains a cofinal partially ordered set (in the notation of Exercise 1.2.4, if $(A, \leq)$ is the preordered set, it suffices to take a complete set of representatives of the elements of A modulo the equivalence relation $\sim$). Thus, this Exercise 4.6.8 shows once more that there is no difference in constructing direct systems over preordered sets or partial ordered sets.

Also notice that any countable directed set contains a cofinal set isomorphic to $(\mathbb{N}, \leq)$.

Direct Limit as a Functor

Fix a ring R and a set I. As we have anticipated in Remark 4.6.1, the direct systems of right R-modules over the directed set I form a category $(\text{Mod}-R)^I$. We will now be more precise and see that direct limit can be viewed as a functor.

We now define the category $(\text{Mod}-R)^I$ of direct systems. The preordered directed set $(I, \leq)$ and the ring R are fixed once for all. The objects of $(\text{Mod}-R)^I$ are the direct systems $\mathbb{M} = (M_i, \mu_{ij})$ of right R-modules over I. If $\mathbb{M} := (M_i, \mu_{ij})$, $\mathbb{N} = (N_i, \nu_{ij})$ are two objects of $(\text{Mod}-R)^I$, a morphism

$$\Phi \colon \mathbb{M} \to \mathbb{N}$$

in the category $(\text{Mod}-R)^I$ is by definition a family $(\varphi_i)_{i \in I}$ of R-module morphisms $\varphi_i \colon M_i \to N_i$ such that $\nu_{ij} \circ \varphi_i = \varphi_j \circ \mu_{ij}$ for every $i \leq j$, that is, such that the diagram

$$\begin{array}{ccc} M_i & \xrightarrow{\varphi_i} & N_i \\ \mu_{ij} \downarrow & & \downarrow \nu_{ij} \\ M_j & \xrightarrow{\varphi_j} & N_j \end{array}$$

commutes for every $i \leq j$. The composite morphism of $\Phi \colon \mathbb{M} \to \mathbb{N}$ and $\Psi \colon \mathbb{N} \to \mathbb{P}$, where $\Phi = (\varphi_i)_{i \in I}$ and $\Psi = (\psi_i)_{i \in I}$ is $\Psi \circ \Phi = (\psi \circ \varphi_i)_{i \in I}$. The identity of $\mathbb{M} = (M_i, \mu_{ij})$ is $1_\mathbb{M} = (1_{M_i})_{i \in I}$. Notice that $\text{Hom}_{(\text{Mod}-R)^I}(\mathbb{M}, \mathbb{N})$ is essentially a subset of $\prod_{i \in I} \text{Hom}_{\text{Mod}-R}(M_i, N_i)$.

Let $\Phi\colon \mathbb{M} \to \mathbb{N}$ be a morphism between two direct systems $\mathbb{M} := (M_i, \mu_{ij})$ and $\mathbb{N} = (N_i, \nu_{ij})$, and let $M := \varinjlim M_i$ and $N := \varinjlim N_i$ be their direct limits. Then the morphisms $\nu_i \circ \varphi_i \colon M_i \to N$ are compatible in the sense that for every $i \leq j$ we have $(\nu_j \circ \varphi_j) \circ \mu_{ij} = \nu_j \circ \nu_{ij} \circ \varphi_i = \nu_i \circ \varphi_i$. By the universal property of (M, μ_i), there exists a unique $\varphi\colon M \to N$ such that $\varphi \circ \mu_i = \nu_i \circ \varphi_i$ for every $i \in I$. It follows easily that:

Proposition 4.6.9
Direct limit is a covariant functor

$$\varinjlim \colon (\mathrm{Mod}{-}R)^I \to \mathrm{Mod}{-}R.$$

We will see a proof of Proposition 4.6.9 more in detail in Exercise 5.6.8. Notice that if $\mathbb{M}$ and $\mathbb{N}$ are viewed as two functors $I \to \mathrm{Mod}{-}R$, then the morphism $\Phi\colon \mathbb{M} \to \mathbb{N}$ turns out to be a natural transformation between the two functors.

Example 4.6.10 (Direct Limits Commute with Tensor Products)
Let $\mathbb{M} = (M_i, \mu_{ij})$ be a direct system of right R-modules over a directed set I and (M, μ_i) be the direct limit of $\mathbb{M}$. Let N be a left R-module. Then $(M_i \otimes_R N, \mu_{ij} \otimes 1_N)$ is a direct system of abelian groups. Let $P := \varinjlim(M_i \otimes_R N)$ be its direct limit. More precisely, this direct limit is the pair (P, π_i) where, for each $i \in I$, $\pi_i \colon M_i \otimes_R N \to P$ is a group morphism with $\pi_i = \pi_j \circ (\mu_{ij} \otimes 1_N)$ for every $i \leq j$. We have a family of group morphisms

$$\mu_i \otimes 1_N \colon M_i \otimes_R N \to M \otimes_R N, \quad i \in I.$$

By the Universal property of direct limit, this family of group morphisms defines a unique group morphism $\psi\colon P \to M \otimes_R N$ with the property that $\psi \circ \pi_i = \mu_i \otimes 1_N$ for every $i \in I$. We will now show that the group morphism ψ has an inverse, hence it is an isomorphism, so that

$$\varinjlim(M_i \otimes_R N) \cong (\varinjlim M_i) \otimes_R N.$$

Let $\beta\colon M \times N \longrightarrow P$ be defined as follows. If $x \in M$ and $y \in N$, there exist $i \in I$ and $x_i \in M_i$ such that $x = \mu_i(x_i)$. Set $\beta(x, y) = \beta(\mu_i(x_i), y) := \pi_i(x_i \otimes y)$. In order to show that β is well defined, assume that $j \in I$ and $x_j \in M_j$ also have the property that $x = \mu_j(x_j)$. Then there exists $k \in I$ with $k \geq i$ and $k \geq j$. The equality $\mu_i(x_i) = \mu_j(x_j)$ can be rewritten as $\mu_k \mu_{ik}(x_i) = \mu_k \mu_{jk}(x_j)$, so that $\mu_k(\mu_{ik}(x_i) - \mu_{jk}(x_j)) = 0$. By Lemma 4.6.4, there exists $l \geq k$ such that $\mu_{kl}(\mu_{ik}(x_i) - \mu_{jk}(x_j)) = 0$. Thus $\mu_{il}(x_i) = \mu_{jl}(x_j)$, so that $(\mu_{il} \otimes 1_N)(x_i \otimes y) = (\mu_{il} \otimes 1_N)(x_j \otimes y)$. It follows that $\pi_l \circ (\mu_{il} \otimes 1_N)(x_i \otimes y) = \pi_l \circ (\mu_{il} \otimes 1_N)(x_j \otimes y)$, i.e., $\pi_i(x_i \otimes y) = \pi_i(x_j \otimes y)$. This proves that β is well defined. We leave to the reader the verification that β is balanced, so that β induces a group homomorphism $\widetilde{\beta}\colon M \otimes_R N \to P$, and that $\widetilde{\beta}$ is the inverse of ψ.

Exercise 4.6.11
Consider the partially ordered set $(\mathbb{N}, \leq)$ of nonnegative integers. For every $i \in \mathbb{N}$, consider the right $\mathbb{Z}$-modules $M_i = \mathbb{Z}_\mathbb{Z}$. For every $i, j \in \mathbb{N}$ with $i \leq j$, let $\mu_{ij} \colon M_i \to M_j$ be the morphisms of $\mathbb{Z}$ into $\mathbb{Z}$ defined by $\mu_{ij}(x) = 2^{j-i}x$ for all $x \in \mathbb{Z} = M_i$. Show that the direct limit $\varinjlim M_i$ is isomorphic to $\mathbb{Z}_2 := \{ \frac{a}{2^n} \mid a \in \mathbb{Z}, n \in \mathbb{N} \}$.

4.7 Direct Limits of Rings

Exercise 4.6.12
Let $(I, \leq)$ be a preordered set with a greatest element i_0 (so that, in particular, I is directed). Show that (M_{i_0}, μ_{ii_0}) is the direct limit of the direct system $\mathbb{M} = (M_i, \mu_{ij})$.

Exercise 4.6.13
Fix a prime p. Consider the partially ordered set $(\mathbb{N}, \leq)$ of nonnegative integers. For every $i \in \mathbb{N}$, let M_i be the cyclic group $\mathbb{Z}/p^i\mathbb{Z}$, that is, cyclic $\mathbb{Z}$-module with p^i elements. For $i, j \in \mathbb{N}$ and $i \leq j$, let $\mu_{ij}: M_i = \mathbb{Z}/p^i\mathbb{Z} \to M_j = \mathbb{Z}/p^j\mathbb{Z}$ be the morphisms defined by $\mu_{ij}(x + p^i\mathbb{Z}) = p^{j-i}x + p^j\mathbb{Z}$ for every $x \in \mathbb{Z} = M_i$. Show that the μ_{ij} are well defined group homomorphisms and that the direct limit $\varinjlim \mathbb{Z}/p^i\mathbb{Z}$ is isomorphic to the Prüfer group $\mathbb{Z}(p^\infty)$.

Exercise 4.6.14
Consider an arbitrary family M_λ, $\lambda \in \Lambda$, of right R-modules. Let $\mathcal{P}_f(\Lambda)$ be the set of all finite subsets F of Λ. Partially order $\mathcal{P}_f(\Lambda)$ by set inclusion, so that $\mathcal{P}_f(\Lambda)$ turns out to be a directed set. Show that the direct sum of the family M_λ, $\lambda \in \Lambda$, can be viewed as the direct limit over the directed set $\mathcal{P}_f(\Lambda)$ of the direct sums of the modules M_λ, $\lambda \in F$, that is, $\bigoplus_{\lambda \in \Lambda} M_\lambda$ is the direct limit

$$\varinjlim \oplus_{\lambda \in F} M_\lambda,$$

where F ranges in $\mathcal{P}_f(\Lambda)$.

4.7 Direct Limits of Rings

Let $(R_i)_{i \in I}$ be a family of rings indexed on a directed set I. For every $i, j \in I, i \leq j$, let $\mu_{ij}: R_i \to R_j$ be a ring morphism, such that if $i \leq j \leq k$ then $\mu_{jk} \circ \mu_{ij} = \mu_{ik}$, and $\mu_{ii} = \iota_{R_i}$ for every $i \in I$. This is a *direct system of rings*. The association $I \to \mathbf{Rings}$ can be also seen a functor. Let R be the direct limit of the $\mathbb{Z}$-modules R_i.

Now we define a multiplication on R. If $x, y \in R$, then $x = \mu_i(x_i)$, $y = \mu_j(y_j)$ for suitable $i, j \in I$, $x_i \in R_i$, $y_j \in R_j$. Let $k \in I$ be such that $i, j \leq k$. Define

$$x \cdot y := \mu_k(\mu_{ik}(x_i)\mu_{jk}(y_j)).$$

This operation turns out to be well defined and R becomes a ring, whose identity is $\mu_i(1_{R_i})$ for (every) $i \in I$.

Exercise 4.7.1
Show that for every ring S and ring morphisms $\alpha_i: R_i \to S$ such that $\alpha_j \circ \mu_{ij} = \alpha_i$ for every $i \leq j$, there exists a unique ring morphism $\alpha: R \to S$ such that $\alpha \circ \mu_i = \alpha_i$.

Exercise 4.7.2
If R_i is commutative for every $i \in I$, then the direct limit is commutative.

Exercise 4.7.3
If R_i is an integral domain for every $i \in I$, then the direct limit is an integral domain.

Exercise 4.7.4
Let k be a field and X be a set of commuting indeterminates. Let $k[X]$ be the ring of polynomials. Let $\mathcal{P}_f(X)$ be the set of all finite subsets of X, so that $\mathcal{P}_f(X)$ is a directed partially ordered set under set inclusion. Show that the direct limit of the rings $k[x_t \mid t \in F]$, where F ranges in $\mathcal{P}_f(X)$, is $k[X]$.

4.8 Flat Modules

Definition 4.8.1

A right R-module M_R is *flat* if the functor

$$M_R \otimes_R -: R-\text{Mod} \to Ab$$

is exact. Equivalently, if for every left R-module monomorphism $\varphi: {}_RA \to {}_RB$, the group homomorphism $1_M \otimes \varphi: M \otimes_R A \to M \otimes_R B$ is also a monomorphism. Similarly, a left R-module ${}_RM$ is *flat* if the functor $- \otimes_R M: \text{Mod}-R \to Ab$ is exact, or, equivalently, if for every right R-module monomorphism $\varphi: A \to B$, the abelian group morphism $\varphi \otimes \iota_M: A_R \otimes_R M \to B \otimes_R M$ is a monomorphism.

Let R be a commutative integral domain. An R-module M_R is *divisible* if for every $r \in R$, $r \neq 0$, and every $x \in M_R$ there exists $y \in M_R$ with $x = ry$. Equivalently, a module M_R is divisible if for every non-zero element $r \in R$, multiplication by r is a surjective homomorphism $\lambda_r: M_R \to M_R$. Notice that we are extending the notion of divisible abelian group to modules over arbitrary integral domains.

Example 4.8.2
We will now show that if R is a commutative integral domain, T is a torsion R-module and D is a divisible R-module, then $T \otimes_R D = 0$.

The elements of $T \otimes_R D$ are finite sums of elements of the form $t \otimes d$ with $t \in T$ and $d \in D$. Now if $t \in T$, there exists a non-zero $r \in R$ such that $tr = 0$. But D is divisible, so that there exists $d' \in D$ with $d = rd'$. Then $t \otimes d = t \otimes rd' = tr \otimes d' = 0 \otimes d' = 0$. Hence $T \otimes_R D = 0$.

Example 4.8.3
Here is an example of a module that is not flat. Consider the embedding $\varphi: \mathbb{Z} \hookrightarrow \mathbb{Q}$, so that $1_{\mathbb{Z}/2\mathbb{Z}} \otimes \varphi$ is a group homomorphism of $\mathbb{Z}/2\mathbb{Z} \otimes_{\mathbb{Z}} \mathbb{Z}$ into $\mathbb{Z}/2\mathbb{Z} \otimes_{\mathbb{Z}} \mathbb{Q}$. Now $\mathbb{Z}/2\mathbb{Z} \otimes_{\mathbb{Z}} \mathbb{Z} \cong \mathbb{Z}/2\mathbb{Z}$ (in general, $- \otimes_R R: \text{Mod}-R \to Ab$ and the forgetful functor $\text{Mod}-R \to Ab$ are naturally isomorphic, Exercise 2.7.14) and $\mathbb{Z}/2\mathbb{Z} \otimes_{\mathbb{Z}} \mathbb{Q} = 0$ because $\mathbb{Z}/2\mathbb{Z}$ is a torsion $\mathbb{Z}$-module and ${}_{\mathbb{Z}}\mathbb{Q}$ is divisible. Thus the mapping $1_{\mathbb{Z}/2\mathbb{Z}} \circ \varphi: \mathbb{Z}/2\mathbb{Z} \otimes_{\mathbb{Z}} \mathbb{Z} \to \mathbb{Z}/2\mathbb{Z} \otimes_{\mathbb{Z}} \mathbb{Q}$ is a morphism from a non-zero group into a zero group. Thus $1_{\mathbb{Z}/2\mathbb{Z}} \circ \varphi$ is not an injective mapping. Hence $\mathbb{Z}/2\mathbb{Z}$ is not a flat $\mathbb{Z}$-module. We will characterize flat $\mathbb{Z}$-modules, more generally flat modules over principal ideal domains, in Proposition 4.8.7(2).

Example 4.8.4
For any ring R, the right R-module R_R is flat. To see this, recall that $R \otimes_R M \cong M$ for every left R-module ${}_RM$ (Exercise 2.7.14). More precisely, there is a natural isomorphism η of the functor $R \otimes_R -: R-\text{Mod} \to Ab$ into the forgetful functor $F: R-\text{Mod} \to Ab$, defined, for every left module ${}_RM$, by $\eta_M: R \otimes_R M \cong M$, $\eta_M: r \otimes_R x \cong rx$.

4.8 Flat Modules

The next lemma describes the first elementary properties of flat modules. We leave its proof as an exercise.

Lemma 4.8.5
(1) *Every direct sum of flat modules is flat.*
(2) *Every direct summmand of a flat module is flat.*
(3) *Every projective module is flat.*

Proposition 4.8.6
Every direct limit of flat modules over a directed set is flat.

Proof Let I be a directed set and (M_i, μ_{ij}) a direct systems of right R-modules in which every M_i flat. Consider a short exact sequence of left R-modules $0 \longrightarrow {}_R N' \longrightarrow {}_R N \longrightarrow {}_R N'' \longrightarrow 0$. Then $0 \longrightarrow M_i \otimes N' \longrightarrow M_i \otimes N \longrightarrow M_i \otimes N'' \longrightarrow 0$ is a short exact sequence of abelian groups, so that $0 \longrightarrow (M_i \otimes N') \longrightarrow (M_i \otimes N) \longrightarrow (M_i \otimes N'') \longrightarrow 0$ is a short exact sequence in $\mathbf{Ab}^I$. Then $0 \longrightarrow \varinjlim(M_i \otimes N') \longrightarrow \varinjlim(M_i \otimes N) \longrightarrow \varinjlim(M_i \otimes N'') \longrightarrow 0$ is a short exact sequence of abelian groups. Therefore $0 \longrightarrow M \otimes N' \longrightarrow M \otimes N \longrightarrow M \otimes N'' \longrightarrow 0$ is a short exact sequence of abelian groups, i.e. M_R is flat. □

According to a famous theorem due to Lazard [20], a module is flat if and only if it is a direct limit of free modules, if and only if it is a direct limit of projective modules. Notice that we have proved that every direct limit of free modules is flat.

It is possible to prove that, for a ring R, every right R-module is flat if and only if every left R-module is flat, if and only if R is *von Neumann regular*, that is, for every $x \in R$ there exists $y \in R$ such that $x = xyx$.

Recall that every finitely generated module over a commutative principal ideal domain R is a finite direct sums of cyclic modules (Elementary divisors theorem). A cyclic module R/rR is torsion for $r \neq 0$ and is torsion-free for $r = 0$. It follows that a finitely generated module over a commutative principal ideal domain is torsion-free if and only if it is free.

Proposition 4.8.7
(1) *Every flat module over a commutative integral domain R is torsion-free.*
(2) *If R is a commutative principal ideal domain, an R-module is flat if and only if it is torsion-free.*

Proof

(1) Let M_R be a flat module over the commutative integral domain R. Fix a non-zero element $r \in R$. Then multiplication by r is an R-module morphism $\lambda'_r : {}_R R \to {}_R R$, which is injective because r is non-zero. Since M_R is flat, the abelian group morphism $\iota_M \otimes \lambda'_r : M \otimes_R R \to M \otimes_R R$ is injective. Now we have a diagram

$$\begin{array}{ccc} M \otimes_R R & \xrightarrow{\iota_M \otimes \lambda'_r} & M \otimes_R R \\ \eta_M \downarrow & & \downarrow \eta_M \\ M & \xrightarrow{\lambda_r} & M, \end{array}$$

which is commutative, because $\eta_M \circ (\iota_M \otimes \lambda'_r)(x \otimes a) = \eta_M \circ (x \otimes ar) = xar$ and $\lambda_r \circ \eta_M(x \otimes a) = \lambda_r(xa) = xar$ for every $x \in M, a \in R$. Since the upper arrow $\iota_M \otimes \lambda'_r$ is a monomorphism and the two vertical arrows η_M are isomorphisms, it follows that the lower arrow λ_r is a monomorphism. This shows that ${}_R M$ is torsion-free.

(2) Let M_R be a torsion-free module over the commutative principal ideal domain R. Then M_R is the union of its finitely generated submodules. Notice that the set $\mathcal{F}$ of all finitely generated submodules of M_R is a directed set with respect to the partial order $\subseteq$, so that M_R is the direct limit of its finitely generated submodules. Every finitely generated submodule of the torsion-free module M_R is torsion-free, hence it is free because R is a principal ideal ring. Thus M_R is a direct limit of free modules, hence M_R is a flat module (Proposition 4.8.6). □

Exercise 4.8.8

Let R be a commutative integral domain and Q its field of fractions. Show that Q is a flat R-module. [*Hint.* The set of cyclic R-submodules of Q is a directed partially ordered set under inclusion. Every non-zero cyclic R-submodule of Q is isomorphic to R, hence is flat. Thus Q, union of its cyclic R-submodules, is a direct limit of flat R-modules.]

4.9 Finitely Presented Modules

Definition 4.9.1

A module M_R is *finitely presented* if it is finitely generated and, for every epimorphism $\varphi : F_R \to M_R$ with F_R a finitely generated free R-module, the kernel of φ is finitely generated.

Example 4.9.2

A ring R is right noetherian if and only if every finitely generated right R-module is finitely presented.

4.9 Finitely Presented Modules

Proof

($\Rightarrow$) Let R be right noetherian. Take M_R finitely generated and an epimorphism $\varphi\colon F_R \to M$ with F free and finitely generated. F_R is noetherian because R_R is, so $\ker(\varphi) \leq F$ is finitely generated.

($\Leftarrow$) Let $I \leq R_R$. Then R/I is a finitely generated right R-module, so it is finitely presented. So the natural projection $\pi\colon R \to R/I$ has finitely generated kernel. But the kernel is exactly I. We have proved that every right ideal of R is finitely generated, i.e., R is right noetherian. $\square$

Example 4.9.3

Every finitely generated projective module is finitely presented.

Proof Let P_R be projective and finitely generated. Take an epimorphism $\varphi\colon F_R \to P_R$, where F is finitely generated and free. Since P is projective, $\ker(\varphi) \leq_\oplus F$, so $\ker(\varphi)$ is an homomorphic image of F, hence finitely generated. $\square$

Lemma 4.9.4 (Schanuel's Lemma)

Let

$$0 \longrightarrow K \longrightarrow P \xrightarrow{q} M \longrightarrow 0$$

$$0 \longrightarrow K' \longrightarrow P' \xrightarrow{q'} M \longrightarrow 0$$

be two short exact sequences of right R modules with P, P' projective modules. Then $P' \oplus K \cong P \oplus K'$.

Dually, let

$$0 \longrightarrow M \xrightarrow{f} E \longrightarrow C \longrightarrow 0$$

$$0 \longrightarrow M \xrightarrow{f'} E' \longrightarrow C' \longrightarrow 0$$

be two short exact sequences of right R-modules with E, E' injective modules. Then $E \oplus C' \cong E' \oplus C$.

Proof Consider
$$Q := \{(p, p') \in P \oplus P' \mid q(p) = q'(p')\} \leq P \oplus P'.$$

Let $\pi_1 \colon Q \to P$ be the projection on the first factor. The mapping π_1 is onto because q' is onto. Since P is projective, $Q \cong P \oplus \ker(\pi_1)$. But $\ker(\pi_1) = 0 \oplus \ker(q') \cong \ker(q') = K'$, so $Q \cong P \oplus K'$. Similarly, using π_2, we get that $Q \cong P' \oplus K$. The result follows.

For the injective case proceed similarly with $Q := \mathrm{coker}((f, f'))$. □

> **Corollary 4.9.5**
>
> A module M_R is finitely presented if and only if it is isomorphic to P_R/S where P_R is finitely generated and projective, and S is a finitely generated submodule of P_R.

Proof
($\Rightarrow$) If M_R is finitely presented then it is finitely generated, so there is an epimorphism $\varphi \colon F \to M$ with F_R finitely generated and free. Then $\ker(\varphi)$ is finitely generated and $M \cong F/\ker(\varphi)$. Moreover, F is projective, because it is free.

($\Leftarrow$) Assume that $M \cong P/S$ as in the statement. Since M is a homomorphic image of the finitely generated module P, it is finitely generated. Take any epimorphism $\varphi \colon F \to M$ with F_R finitely generated and free. Apply Schanuel's lemma to the two short exact sequences

$$0 \longrightarrow \ker(\varphi) \longrightarrow F \longrightarrow M \longrightarrow 0$$

$$0 \longrightarrow S \longrightarrow P \longrightarrow M \longrightarrow 0$$

getting that $S \oplus F \cong \ker(\varphi) \oplus P$. Since S and F are finitely generated, so is $S \oplus F$. Therefore $\ker(\varphi)$, which is a homomorphic image of $S \oplus F$, is also finitely generated.

□

Thus we could have defined a finitely presented module as a right R-module M_R for which there exists a short exact sequence

$$R_R^m \longrightarrow R_R^n \longrightarrow M_R \longrightarrow 0$$

for suitable integers $n, m \geq 0$.

4.9 Finitely Presented Modules

> **Theorem 4.9.6**
> *Every right module is a direct limit of finitely presented modules.*

Proof Fix M_R. Then $M \cong F/S$ for a suitable free R-module F_R and $S \leq F$. Let

$$I := \{(A, B) \mid A \leq F, B \leq A \cap S, \ A \ f.g. \ and \ free, \ B \ f.g.\}.$$

Partially order I by "double inclusion". We now show that $(I, \leq)$ is directed. If $(A, B), (A', B') \in I$ then there is a finitely generated and free submodule A'' of F with $A'' \supseteq A + A'$, so that $(A'', B + B') \geq (A, B), (A', B')$. For every $(A, B) \in I$ define

$$S_{(A,B)} := B, \qquad F_{(A,B)} := A, \qquad M_{(A,B)} := A/B.$$

Then we have a short exact sequence

$$0 \longrightarrow S_{(A,B)} \longrightarrow F_{(A,B)} \longrightarrow M_{(A,B)} \longrightarrow 0$$

If $(A, B), (A', B') \in I$ and $(A, B) \leq (A', B')$, then we have a commutative diagram with exact rows

$$\begin{array}{ccccccccc}
0 & \longrightarrow & S_{(A,B)} & \longrightarrow & F_{(A,B)} & \longrightarrow & M_{(A,B)} & \longrightarrow & 0 \\
& & \downarrow & & \downarrow & & \downarrow & & \\
0 & \longrightarrow & S_{(A',B')} & \longrightarrow & F_{(A',B')} & \longrightarrow & M_{(A',B')} & \longrightarrow & 0,
\end{array}$$

where the first two vertical arrows are inclusions, and the third follows by requiring commutativity.

It's easy to see that it yields a short exact sequence in $(\mathrm{Mod}-R)^I$

$$0 \longrightarrow \mathbb{S} \longrightarrow \mathbb{F} \longrightarrow \mathbb{M} \longrightarrow 0$$

Apply the exact functor $\varinjlim$ getting a short exact sequence

$$0 \longrightarrow \varinjlim(S_{(A,B)}) \longrightarrow \varinjlim(F_{(A,B)}) \longrightarrow \varinjlim(M_{(A,B)}) \longrightarrow 0$$

Then

$$\varinjlim(M_{(A,B)}) \cong \varinjlim(F_{(A,B)})/\varinjlim(S_{(A,B)}) \cong$$
$$\cong \cup\{A \mid (A,B) \in I\}/\cup\{B \mid (A,B) \in I\} =$$
$$= F/S \cong M.$$

We are done because the $M_{(A,B)}$ are finitely presented. □

4.10 Inverse Limits

Let $(I, \leq)$ be a preordered set. An *inverse system* (or *projective system*) (M_i, μ_{ji}) of right R-modules over I consists of a right R-module M_i for every $i \in I$ and an R-module morphism $\mu_{ji}: M_j \to M_i$ for every $i \leq j$, such that:

(a) if $i \leq j \leq k$, then $\mu_{ki} = \mu_{ji} \circ \mu_{kj}$, and
(b) $\mu_{ii} = 1_{M_i}$ for every $i \in I$.

In other words, if we view at I as a category as we have done in Example 1.2.2(7), an inverse system is exactly a contravariant functor $I \to \text{Mod}-R$.

The *inverse limit* (or *projective limit*, or simply *limit*) of the inverse system (M_i, μ_{ji}) is the submodule

$$M = \varprojlim M_i := \{(x_i)_{i \in I} \in \prod_{i \in I} M_i \mid \mu_{ji}(x_j) = x_i \text{ for every } i \leq j\}$$

of the direct product $\prod_{i \in I} M_i$ with canonical morphisms $\mu_j: \varprojlim M_i \to M_j$, which are the restrictions to the module M of the canonical projections. It is easily seen that $\mu_{ji} \circ \mu_j = \mu_i$ for every $i \leq j$.

Exercise 4.10.1
Prove that the inverse limit M is characterized, up to isomorphism, by the following *Universal Property of Inverse Limit*:

Let I be a directed preordered set, $\mathbb{M} = (M_i, \mu_{ji})$ an inverse system of right modules over a ring R, and (M, μ_i) its inverse limit. Let N_R be a right R-module and let $\alpha_i: N \to M_i, i \in I$, be a family of right R-module morphisms such that $\mu_{ji} \circ \alpha_j = \alpha_i$ for every $i \leq j$. Then there is a unique morphism $\alpha: N \to M$ such that $\alpha_i = \mu_i \circ \alpha$ for every $i \in I$.

There is a canonical isomorphism

$$\varprojlim \text{Hom}(Y, M_i) \cong \text{Hom}(Y, \varprojlim M_i).$$

4.10 Inverse Limits

Example 4.10.2
Take $I := (\mathbb{N}, \leq)$. It is a directed set. For $i \in I$ take $M_i := \mathbb{Z}_\mathbb{Z}$. For $i \leq j$ let $\mu_{ji}: M_j \to M_i$ be the multiplication by 2^{j-i}. $(\mathbb{Z}, \mu_{ji})$ is an inverse system over $\mathbb{N}$ and the inverse limit is 0.

Exercise 4.10.3
Also in this case, the fact that the preordered set is directed is not strictly necessary. Show that if $(I, =)$ is a set I preordered by the equality relation $=$, like in Example 4.6.3, then the inverse limit of a direct system $\mathbb{M} = (M_i, \mu_{ii})$ is the direct product $\prod_{i \in I} M_i$.

Notice that there is no obvious dualization of the fact that in a direct limit every element is the image of some M_i and that if an element is 0 at infinity, then it is 0 at a "finite step".

If $\{M_i \mid i \in I\}$ is a family of submodules of a module M_R, and the family is directed with respect to inverse inclusion, that is, for every $i, j \in I$ there exists $k \in I$ such that $M_k \subseteq M_i \cap M_j$, let $\mu_{ji}: M_j \to M_i$ be the inclusion for every $i, j \in I$ with $M_j \subseteq M_i$. Then the inverse limit of the inverse system (M_i, μ_{ji}) is (isomorphic to) the intersection of the modules M_i.

> **Theorem 4.10.4** ([4, Theorem 2])
> *If R is a ring and M_R is a right R-module, then M_R can be written as the intersection of a downward directed system of injective submodules of an injective module.*

Thus every module is an inverse limit of injective modules. This contrasts with the fact that direct limits of projective modules are exactly flat modules (page 165).

We can define the category of inverse systems in the natural way, as for direct systems, (one must define what a morphism between two inverse system is). Inverse limit turns out to be a covariant functor of this category into Mod$-R$. This functor (which is additive in the sense of Definition 5.3.5) is left exact, but not exact in general, as the following exercise shows.

Exercise 4.10.5
Let k be a field, and let $\{v_n \mid n \geq 0\}$ be a basis of a countably dimensional vector space over k. For every $n \geq 0$, let V_n be the vector space with basis $\{v_n, v_{n+1}, v_{n+2}, \dots\}$, so that we have a strictly descending chain $V_0 \supset V_1 \supset V_0 \supset \dots$ of vector spaces. Then, for every pair $n \leq m$ of nonnegative integers, there is a commutative diagram

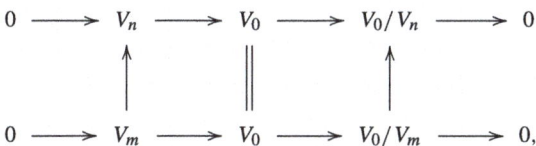

where all the mappings are the canonical ones (inclusions, the identity, or the canonical projections). Deduce from this that the functor inverse limit is not right exact.

Additive Categories, Abelian Categories 5

5.1 Monomorphisms, Epimorphisms, Subobjects, Quotient Objects

Recall that a morphism $f\colon C \to C'$ in a category $\mathcal{C}$ is an *isomorphism* if there exists a morphism $g\colon C' \to C$ such that $g \circ f = 1_C$ and $f \circ g = 1_{C'}$.

Definition 5.1.1

Let $\mathcal{C}$ be a category and $f\colon C \to C'$ a morphism. We say that:

(a) f is a *monomorphism* if, for any object B in $\mathcal{C}$ and any two morphisms $g, h\colon B \to C$, $f \circ g = f \circ h$ implies $g = h$.
(b) f is an *epimorphism* if, for any object D in $\mathcal{C}$ and any two morphisms $g, h\colon C' \to D$, $g \circ f = h \circ f$ implies $g = h$.

Example 5.1.2
Let R be a ring and let $\mathcal{C}$ be the category defined as follows. $\mathcal{C}$ has a unique object $*$, i.e., $\mathrm{Ob}(\mathcal{C}) := \{*\}$, $\mathrm{Hom}_{\mathcal{C}}(*, *) := R$, and $\circ\colon \mathrm{Hom}_{\mathcal{C}}(*, *) \times \mathrm{Hom}_{\mathcal{C}}(*, *) \to \mathrm{Hom}_{\mathcal{C}}(*, *)$, $(r, s) \mapsto rs$, that is, composition in $\mathcal{C}$ is exactly the multiplication in R. The isomorphisms in $\mathcal{C}$ are the units of R. The monomorphisms in $\mathcal{C}$ are the elements of R that are not left zero-divisors. The epimorphisms in $\mathcal{C}$ are the elements of R that are not right zero-divisors.

Exercise 5.1.3
A *concrete category* is a pair $(\mathcal{C}, F)$, where $\mathcal{C}$ is a category and $F\colon \mathcal{C} \to \mathbf{Set}$ is a faithful functor. In a concrete category we can identify every object $C \in \mathrm{Ob}(\mathcal{C})$ with the set $F(C)$ and every morphism f with the mapping $F(f)$. Show that if f is a morphism and the mapping $F(f)$ is injective, then f is a monomorphism. Dually, if $F(f)$ is surjective, then f is an epimorphism. Also, if f is an isomorphism, then $F(f)$ is a bijection.

Exercise 5.1.4
Show that any isomorphism is both a monomorphism and an epimorphism.

Example 5.1.5
In the category **Rng**, the inclusion $\varepsilon \colon \mathbb{Z} \to \mathbb{Q}$ is both a monomorphism and an epimorphism, but not an isomorphism.

It is a monomorphism, because it is a monomorphism in **Set**, hence a fortiori in **Rng**. In order to show that ε is an epimorphism, we must prove that for any two ring morphisms f, g of $\mathbb{Q}$ into a ring R, if for their restrictions we have that $f|_{\mathbb{Z}} = g|_{\mathbb{Z}}$, then $f = g$. Now an arbitrary element of $\mathbb{Q}$ is a fraction $\frac{m}{n}$, with m, n integers and n non-zero. The ring morphisms f, g are multiplicative monoid morphisms of $(\mathbb{Q}, \cdot)$ into $(R, \cdot)$, hence induce by restriction two multiplicative group morphisms of $(U(\mathbb{Q}), \cdot)$ into $(U(R), \cdot)$ between the groups of invertible elements. In particular, if $u \in \mathbb{Q}$ is invertible in $\mathbb{Q}$, then $f(u)$ and $g(u)$ are invertible in R and $f(u^{-1}) = f(u)^{-1}$, $g(u^{-1}) = g(u)^{-1}$. Thus $f|_{\mathbb{Z}} = g|_{\mathbb{Z}}$ implies that, for every $\frac{m}{n} \in \mathbb{Q}$,

$$f(\tfrac{m}{n}) = f(mn^{-1}) = f(m)f(n^{-1}) = f(m)f(n)^{-1} =$$
$$= g(m)g(n)^{-1} = g(mn^{-1}) = g(\tfrac{m}{n}),$$

that is, $f = g$.

Finally, ε is not an isomorphism, because if there exists a ring morphism $\varphi \colon \mathbb{Q} \to \mathbb{Z}$ with $\varphi\varepsilon = 1_{\mathbb{Z}}$ and $\varepsilon\varphi = 1_{\mathbb{Q}}$, then the mapping ε is a bijection, which is not.

Let $\mathcal{C}$ be a category and let A an object of $\mathcal{C}$. Consider the class

$$\mathcal{M}_A := \{ f \colon B \to A \mid B \in \mathrm{Ob}(\mathcal{C}) \text{ and } f \text{ is a monomorphism} \}.$$

On $\mathcal{M}_A$ define a relation $\leq$ setting, for every $f \colon B \to A$, $f' \colon B' \to A$ in $\mathcal{M}_A$, $f \leq f'$ if there exists a morphism $g \colon B \to B'$ with $f = f' \circ g$. Notice that, in this case, such a g is necessarily a monomorphism. The relation $\leq$ on the class $\mathcal{M}_A$ is reflexive and transitive. Define an equivalence relation $\sim$ on $\mathcal{M}_A$ setting $f \sim f'$ if $f \leq f'$ and $f' \leq f$. Then $\sim$ turns out to be an equivalence on $\mathcal{M}_A$. The equivalence classes $[f]_\sim$ are called *subobjects* of A. The quotient class $\mathcal{M}_A/\sim$ is called the *class of all subobjects* of A.

Dually for quotient objects of A. One considers the class $\mathcal{E}_A := \{ f \colon A \to B \mid B \in \mathrm{Ob}(\mathcal{C}) \text{ and } f \text{ is an epimorphism} \}$. The quotient objects of A are isomorphism classes of epimorphisms in $\mathcal{E}_A$ modulo a suitable equivalence relation $\sim$.

Exercise 5.1.6
Show that in the category **Set**, if A is a set, there is a one-to-one correspondence between the class $\mathcal{M}_A/\sim$ of all subobjects of A in **Set** and the set $\mathcal{P}(A)$ of all subsets of A. Dually, there is a one-to-one correspondence between the class $\mathcal{E}_A/\sim$ of all quotient objects of A in **Set** and the set of all equivalence relations on the set A.

5.2 Products and Coproducts

Definition 5.2.1

Let $\mathcal{C}$ be a category and A and B be objects of $\mathcal{C}$. A *product* $(A \prod B, \pi_A, \pi_B)$ of A and B in $\mathcal{C}$ consists of an object $A \prod B$ of $\mathcal{C}$ and morphisms $\pi_A \colon A \prod B \to A$ and $\pi_B \colon A \prod B \to B$ such that for any pair of morphisms $f \colon C \to A$, $g \colon C \to$

5.2 Products and Coproducts

B, where C is any object of $\mathcal{C}$, there is a unique morphism $h\colon C \to A \prod B$ such that $\pi_A \circ h = f$ and $\pi_B \circ h = g$. Equivalently,

$$\mathrm{Hom}(C, A \prod B) \to \mathrm{Hom}(C, A) \times \mathrm{Hom}(C, B), \ h \mapsto (\pi_A \circ h, \pi_B \circ h)$$

is a bijection for every $C \in \mathrm{Ob}(\mathcal{C})$.

Thus a product of two objects A and B is the solution of a universal problem. It does not necessarily exist for any two objects A, B in a category $\mathcal{C}$, but when it exists it is always unique up to isomorphism in $\mathcal{C}$.

Definition 5.2.2

Let $\mathcal{C}$ be a category and A, B objects of $\mathcal{C}$. A *coproduct* $(A \coprod B, \varepsilon_A, \varepsilon_B)$ of A and B consists of an object $A \coprod B$ of $\mathcal{C}$ and morphisms $\varepsilon_A \colon A \to A \coprod B$ and $\varepsilon_B \colon B \to A \coprod B$ such that for any pair of morphisms $f \colon A \to C$, $g \colon B \to C$, where C is any object of $\mathcal{C}$, there is a unique morphism $h \colon A \coprod B \to C$ such that $h \circ \varepsilon_A = f$ and $h \circ \varepsilon_B = g$. Equivalently,

$$\mathrm{Hom}(A \coprod B, C) \to \mathrm{Hom}(A, C) \times \mathrm{Hom}(B, C), \ h \mapsto (h \circ \varepsilon_A, h \circ \varepsilon_B)$$

is a bijection for every $C \in \mathrm{Ob}(\mathcal{C})$.

When a coproduct $A \coprod B$ of two objects A and B exists in a category $\mathcal{C}$, it is unique up to isomorphism in $\mathcal{C}$.

Example 5.2.3
Consider the category **Set**. The product of $X, Y \in \mathrm{Ob}(\mathbf{Set})$ $(X \times Y, \pi_X, \pi_Y)$, where $X \times Y$ is the cartesian product and π_X, π_Y are the canonical projections. The coproduct is $(X \dot{\cup} Y, \varepsilon_X, \varepsilon_Y)$, where $X \dot{\cup} Y$ is the disjoint union and $\varepsilon_X, \varepsilon_Y$ are the inclusions.

Example 5.2.4
In the category **Ab**, the product of the abelian groups G and H is $(G \oplus H, \pi_G, \pi_H)$, the coproduct is $(G \oplus H, \varepsilon_G, \varepsilon_H)$.

Example 5.2.5
In the category **Grp** of all groups, the product of two groups G and H is the direct product (cartesian product) $G \times H$ with the canonical projection $\pi_G \colon G \times H \to G$ and $\pi_H \colon G \times H \to H$. The coproduct is $(G * H, \varepsilon_G, \varepsilon_H)$ where $G * H$ is the *free product* of G and H, which is the group defined as follows. Let $G \dot{\cup} H$ be the disjoint union of the sets G and H, and let F be the free group on the set $G \dot{\cup} H$. Let $\varepsilon \colon G \dot{\cup} H \to F$ be the canonical mapping with the universal property of free groups. Set $G * H := F/N$, where N is the normal subgroup of F generated by the set $\{\varepsilon(g)\varepsilon(g')\varepsilon(gg')^{-1}, \ \varepsilon(h)\varepsilon(h')\varepsilon(hh')^{-1} \mid g, g' \in G, \ h, h' \in H\}$. Then $G * H$, called the *free product* of G and H is a group generated by the set $\{\varepsilon(x)N \mid x \in G \dot{\cup} H\}$. Define $\varepsilon_G \colon G \to G * H$, $\varepsilon_H \colon H \to G * H$ by $\varepsilon_G(g) = \varepsilon(g)N$ for every $g \in G$ and $\varepsilon_H(h) = \varepsilon(h)N$ for every $h \in H$. It is easily seen that these two mappings $\varepsilon_G, \varepsilon_H$ are group morphisms. For instance, if $g, g' \in G$, then $\varepsilon_G(g)\varepsilon_G(g') = \varepsilon_G(gg')$, i.e., $\varepsilon(g)N\varepsilon(g')N = \varepsilon(gg')N$, because

$\varepsilon(g)\varepsilon(g')\varepsilon(gg')^{-1} \in N$. It remains to show that the free product with the two morphisms $\varepsilon_G, \varepsilon_H$ satisfies the universal property of coproduct. Let L be a group and $\varphi_G: G \to L, \varphi_H: H \to L$ be two group morphisms. Let $\omega: F \to L$ be the unique group morphism such that $\omega(\varepsilon(g)) = \varphi_G(g)$ for every $g \in G$ and $\omega(\varepsilon(h)) = \varphi_H(h)$ for every $h \in H$. The generators of N are mapped to zero by ω. Hence ω induces a unique group morphism $\widetilde{\omega}: F/N \to L$ such that $\widetilde{\omega}(\varepsilon(g)N) = \varphi_G(g)$ for every $g \in G$ and $\widetilde{\omega}(\varepsilon(h)N) = \varphi_H(h)$ for every $h \in H$. Thus $\widetilde{\omega}\varepsilon_G = \varphi_G$ and $\widetilde{\omega}\varepsilon_H = \varphi_H$. In order to show that $\widetilde{\omega}$ is the unique morphism with this property, assume that also $\omega': F \to L$ is a group morphism such that $\omega'\varepsilon_G = \varphi_G$ and $\omega'\varepsilon_H = \varphi_H$. Then $\omega'(\varepsilon(g)N) = \varphi_G(g)$ for every $g \in G$ and $\omega'(\varepsilon(h)N) = \varphi_H(h)$ for every $h \in H$. Hence ω' and $\widetilde{\omega}$ coincide on the set of generators of F/N. Hence they are equal on F/N.

Example 5.2.6
In the category **Top**, the product of two objects X, Y is the cartesian product $X \times Y$ with the product topology. The coproduct is the disjoint union $X \dot\cup Y$ with the topology in which the open sets are the sets $U \dot\cup V$ in which U, V are open sets in X, Y respectively.

Example 5.2.7
In the categories R-Mod and Mod-R, the product and the coproduct of two modules is their direct sum.

Example 5.2.8
In the category **Rng**, the product of two objects R, S is their direct product (cartesian product) $R \times S$.

The coproduct is $(R * S, \varepsilon_R, \varepsilon_S)$ where $R * S$ is the free product of the rings R and S. It is defined as follows. Let $R \dot\cup H$ be the disjoint union of the sets R and H, and let F be the free ring on the set $R \dot\cup H$, that is, the ring of all non-commutative polynomials with coefficients in $\mathbb{Z}$ in the set of noncommuting indeterminates $R \dot\cup H$. Let $\varepsilon: R \dot\cup H \to F$ be the canonical mapping. Set $R * S := F/I$, where I is the two-sided ideal of F generated by the set $\{\varepsilon(r) + \varepsilon(r') - \varepsilon(r + r'), \varepsilon(r)\varepsilon(r') - \varepsilon(rr'), \varepsilon(s) + \varepsilon(s') - \varepsilon(s + s'), \varepsilon(s)\varepsilon(s') - \varepsilon(ss'), \mid r, r' \in R, s, s' \in S\}$. Then $R * S$ is called the *free product* of R and S. As a ring, it is generated by the set $\{\varepsilon(x) + I \mid x \in R \dot\cup S\}$. Define $\varepsilon_R: R \to R * S$, $\varepsilon_S: S \to R * S$ by $\varepsilon_R(r) = \varepsilon(r) + I$ for every $r \in R$ and $\varepsilon_S(s) = \varepsilon(s) + I$ for every $s \in S$.

Example 5.2.9
Let k be a commutative ring. A commutative k-algebra is a commutative ring R with a morphism $\varphi: k \to R$. Let k-**CommAlg** be the category of all commutative k-algebras. The product of (R, φ_R) and (S, φ_S) is $(R \times S, \varphi_{R \times S})$ where $\varphi_{R \times S}(\lambda) = (\varphi_R(\lambda), \varphi_S(\lambda))$. The coproduct is $(R \otimes_k S, \varphi_{R \otimes S})$, where $\varphi_{R \otimes S}(\lambda) = \varphi_R(\lambda) \otimes_k \varphi_S(\lambda)$.

Example 5.2.10
In the category $\mathcal{C}$ of Example (7) in 1.2.2, constructed from a partially ordered set $(P, \leq)$, the product of two objects $p, q \in P$ is their greatest lower bound $p \wedge q$, and their coproduct is the least upper bound $p \vee q$.

It is possible to define product and coproduct of arbitrary family of objects. Let $\mathcal{C}$ be a category and A_λ be an object of $\mathcal{C}$ for every λ in an index set Λ. A *product*

$$\left(\prod_{\lambda \in \Lambda} A_\lambda, \pi_\lambda\right)$$

of the family $\{A_\lambda \mid \lambda \in \Lambda\}$ consists of an object $\prod_{\lambda \in \Lambda} A_\lambda$ of $\mathcal{C}$ and morphisms $\pi_\mu\colon \prod_{\lambda \in \Lambda} A_\lambda \to A_\mu$ for every $\mu \in \Lambda$ such that for any family of morphisms $f_\mu\colon C \to A_\mu$, $\mu \in \Lambda$, there is a unique morphism $g\colon C \to \prod_{\lambda \in \Lambda} A_\lambda$ such that $\pi_\mu \circ g = f_\mu$ for every $\mu \in \Lambda$. Again, the product of a family of objects does not necessarily exist, but if it exists it is unique up to isomorphism.

Similarly for coproducts. Let $\mathcal{C}$ be a category and $\{A_\lambda \mid \lambda \in \Lambda\}$ a family of objects of $\mathcal{C}$. A *coproduct* $(\coprod_{\lambda \in \Lambda} A_\lambda, \varepsilon_\lambda)$ of the family $\{A_\lambda \mid \lambda \in \Lambda\}$ consists of an object $\coprod_{\lambda \in \Lambda} A_\lambda$ of $\mathcal{C}$ and morphisms $\varepsilon_\mu\colon A_\mu \to \coprod_{\lambda \in \Lambda} A_\lambda$ for each $\mu \in \Lambda$ such that for any object C of $\mathcal{C}$ and any family of morphisms $f_\mu\colon A_\mu \to C$, $\mu \in \Lambda$, there is a unique morphism $g\colon \coprod_{\lambda \in \Lambda} A_\lambda \to C$ such that $g \circ \varepsilon_\mu = f_\mu$ for every $\mu \in \Lambda$.

5.3 Initial Objects, Preadditive Categories

Definition 5.3.1

Let $\mathcal{C}$ be a category. An object Z is called an *initial object* of $\mathcal{C}$ if for every $C \in \mathrm{Ob}(\mathcal{C})$ there is exactly one morphism $Z \to C$. Similarly, Z is called a *terminal object* of $\mathcal{C}$ if for every $C \in \mathrm{Ob}(\mathcal{C})$ there is exactly one morphism $C \to Z$. Finally, Z is called a *null object* (or a *zero object*) if it is both initial and terminal.

Thus an object Z is initial if and only if $\mathrm{Hom}_\mathcal{C}(Z, C)$ has cardinality 1 for every object C, and is terminal if and only if $\mathrm{Hom}_\mathcal{C}(C, Z)$ has cardinality 1 for every C. Obviously, an object is an initial object in a category $\mathcal{C}$ if and only if it is a terminal object in the dual category $\mathcal{C}^{\mathrm{op}}$, and, conversely, an object is a terminal object in $\mathcal{C}$ if and only if it is an initial object in $\mathcal{C}^{\mathrm{op}}$.

Exercise 5.3.2
Show that, in any category, there is at most one initial object up to isomorphism. Similarly for terminal objects

Thus, if a category has an initial (terminal, zero) object, then all its initial (terminal, zero) objects are isomorphic.

Examples 5.3.3
In the category **Set**, the only initial object is the empty set, the terminal objects are the sets of cardinality 1, there is no null object. In the category **Ab**, the initial objects coincide with the terminal objects and coincide with the groups of order 1. In the category **Ring**, the object $\mathbb{Z}$ is an initial object, and the zero ring is a terminal object.

Let $\mathcal{C}$ be a category and let Z be a null object of $\mathcal{C}$. Then there exist exactly one morphism $A \to Z$ and exactly one morphism $Z \to B$ for every pair of objects A, B. Their composition $A \to B$ is called a *zero morphism* of A into B.

Lemma 5.3.4
Let $\mathcal{C}$ be a category with a null object. For every pair of objects A, B of $\mathcal{C}$, there is a unique zero morphism $A \to B$.

Proof Let $\varphi, \varphi' \colon A \to B$ be two zero morphisms, so that there are two null objects Z, Z' such that $\varphi = \beta\alpha$ and $\varphi' = \beta'\alpha'$, where $\alpha \colon A \to Z$, $\beta \colon Z \to B$, $\alpha' \colon A \to Z'$, $\beta' \colon Z' \to B$ are the unique morphisms between the indicated objects. Also, there is a unique morphism $\gamma \colon Z \to Z'$, so that, in particular, $\beta'\gamma = \beta$ and $\gamma\alpha = \alpha'$. Thus $\varphi = \beta\alpha = (\beta'\gamma)\alpha = \beta'(\gamma\alpha) = \beta'\alpha' = \varphi'$.

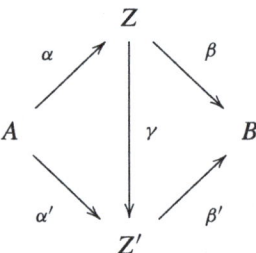

□

The unique zero morphism $A \to B$ will be denoted $\zeta_{A,B}$.

A category $\mathcal{C}$ is a *preadditive* category (or a *$\mathbb{Z}$-category*) if it is possible to assign to the set $\mathrm{Hom}_{\mathcal{C}}(A, B)$ an abelian group structure for every $A, B \in \mathrm{Ob}(\mathcal{C})$ in such a way that the composition $\circ$ is $\mathbb{Z}$-bilinear, that is, for every $f, f' \colon A \to B$ and every $g, g' \colon B \to C$, $g \circ (f + f') = g \circ f + g \circ f'$ and $(g + g') \circ f = g \circ f + g' \circ f$.

Definition 5.3.5

If $\mathcal{C}, \mathcal{D}$ are preadditive categories, a functor

$$F \colon \mathcal{C} \to \mathcal{D}$$

is an *additive* functor if $F(f + f') = F(f) + F(f')$ for every $f, f' \colon A \to B$ in $\mathcal{C}$. Equivalently, if the mapping

$$F_{A,B} \colon \mathrm{Hom}_{\mathcal{C}}(A, B) \to \mathrm{Hom}_{\mathcal{D}}(F(A), F(B))$$

is a group morphism for every $A, B \in \mathrm{Ob}(\mathcal{C})$. Similarly for contravariant functors.

In Example 5.3.12 we will give an example of an isomorphism of categories that is not an isomorphism of preadditive categories.

5.3 Initial Objects, Preadditive Categories

Examples 5.3.6
The categories **Set**, **Rng**, **Grp** and **Top** are not preadditive categories, but a proof of this will be given only after Lemma 5.3.8. The categories **Ab**, Mod-R, R-Mod, **Vect**-K are preadditive. The dual category of a preadditive category is a preadditive category. If R is a ring and $\mathcal{C}$ is the category having exactly one object whose endomorphism set is R, then $\mathcal{C}$ is a preadditive category. For every right module M_R, the functors $\text{Hom}(-, M_R)\colon \text{Mod-}R \to \textbf{Ab}$ and $\text{Hom}(M_R, -)\colon \text{Mod-}R \to \textbf{Ab}$ are additive functors.

If $\mathcal{C}$ is a preadditive category, for any two objects A, B in $\text{Ob}(\mathcal{C})$, the additive abelian group $\text{Hom}_\mathcal{C}(A, B)$ has a zero with respect to the group addition. We will call it the *zero morphism* of A into B and denote it with $0_{A,B}$. We must immediately verify that:

Lemma 5.3.7
If $\mathcal{C}$ is a preadditive category with a null object, the two zero morphisms $\zeta_{A,B}$ and $0_{A,B}$ defined above coincide.

Proof Let Z be a zero object. Composition is a $\mathbb{Z}$-bilinear mapping

$$\text{Hom}(Z, B) \times \text{Hom}(A, Z) \to \text{Hom}(A, B),$$

so that $(0_{\text{Hom}(Z,B)}, 0_{\text{Hom}(A,Z)}) \mapsto 0_{\text{Hom}(Z,B)} \circ 0_{\text{Hom}(A,Z)} = 0_{\text{Hom}(A,B)}$. That is, $\zeta_{A,B} = 0_{A,B}$. $\square$

Lemma 5.3.8
The following conditions are equivalent for an object Z of a preadditive category $\mathcal{C}$:

(a) *Z is an initial object.*
(b) *Z is a terminal object.*
(c) *$1_Z = 0_{Z,Z}$.*
(d) *$|\text{Hom}_\mathcal{C}(Z, Z)| = 1$, that is, $\text{Hom}_\mathcal{C}(Z, Z)$ is the trivial group with one element.*

In other words, in a preadditive category, null objects, initial objects and terminal objects coincide.

Proof (a) $\Rightarrow$ (c) If Z is an initial object, there is a unique morphism $Z \to Z$, so that in particular $1_Z = \zeta_{Z,Z}$. (c) $\Rightarrow$ (d) For every $f \in \text{Hom}_\mathcal{C}(Z, Z)$, one has $f = f \circ 1_Z = f \circ \zeta_{Z,Z} = \zeta_{Z,Z}$. (d) $\Rightarrow$ (a) For every object A, $\text{Hom}_\mathcal{C}(Z, A)$ is a group, hence has at least one element, so that there is a morphism $Z \to A$. Let us show that this morphism is unique if (d) holds. From (d), it follows that $1_Z = 0_{Z,Z}$. Hence, for every $f: Z \to A$, we have $f = f \circ 1_Z = f \circ 0_{Z,Z} = 0_{Z,A}$. Thus $|\operatorname{Hom}_{\mathcal{C}}(Z, A)| = 1$ for every A, that is, Z is an initial object. One proves that (b) is logically equivalent to the other conditions applying (a) $\Leftrightarrow$ (c) $\Leftrightarrow$ (d) to the dual category of $\mathcal{C}$. □

It is now clear that **Set**, **Rng** and **Top** are not preadditive categories. Cf. Example 5.3.3.

Exercise 5.3.9
Show that additive functors send zero morphisms to zero morphisms and null objects to null objects.

Let $\mathcal{C}$ be a preadditive category and let $A_1, A_2 \in \operatorname{Ob}(\mathcal{C})$. A *biproduct* of A_1, A_2 is a quintuple

$$(B, \pi_1, \pi_2, \varepsilon_1, \varepsilon_2),$$

where $B \in \operatorname{Ob}(\mathcal{C})$ and $\pi_1: B \to A_1$, $\pi_2: B \to A_2$, $\varepsilon_1: A_1 \to B$, $\varepsilon_2: A_2 \to B$ are morphisms in $\mathcal{C}$ such that

$$\pi_1 \circ \varepsilon_1 = 1_{A_1}, \quad \pi_2 \circ \varepsilon_2 = 1_{A_2}, \quad \varepsilon_1 \circ \pi_1 + \varepsilon_2 \circ \pi_2 = 1_B$$

Lemma 5.3.10
If $(B, \pi_1, \pi_2, \varepsilon_1, \varepsilon_2)$ is a biproduct, then

$$\pi_2 \circ \varepsilon_1 = 0_{A_1, A_2}, \quad \pi_1 \circ \varepsilon_2 = 0_{A_2, A_1}$$

Proof $\pi_2 \circ \varepsilon_1 = \pi_2 \circ 1_B \circ \varepsilon_1 = \pi_2 \circ (\varepsilon_1 \circ \pi_1 + \varepsilon_2 \circ \pi_2) \circ \varepsilon_1 = \pi_2 \circ \varepsilon_1 \circ \pi_1 \circ \varepsilon_1 + \pi_2 \circ \varepsilon_2 \circ \pi_2 \circ \varepsilon_1 = \pi_2 \circ \varepsilon_1 + \pi_2 \circ \varepsilon_1$, so that $\pi_2 \circ \varepsilon_1 = 0$. Similarly, $\pi_1 \circ \varepsilon_2 = 0$. □

Lemma 5.3.11
Let $\mathcal{C}$ be a preadditive category, and let $A_1, A_2 \in \operatorname{Ob}(\mathcal{C})$. Then

(i) *If $(B, \pi_1, \pi_2, \varepsilon_1, \varepsilon_2)$ is a biproduct of A_1 and A_2, then (B, π_1, π_2) is a product of A_1, A_2.*
(ii) *If (B, π_1, π_2) is a product of A_1 and A_2, then there exist $\varepsilon_1: A_1 \to B$, $\varepsilon_2: A_2 \to B$, morphisms in $\mathcal{C}$, such that $(B, \pi_1, \pi_2, \varepsilon_1, \varepsilon_2)$ is a biproduct of A_1 and A_2.*

5.3 Initial Objects, Preadditive Categories

Proof

(i) Let $(B, \pi_1, \pi_2, \varepsilon_1, \varepsilon_2)$ be a biproduct of A_1 and A_2. We must prove that, for every fixed $C \in \mathrm{Ob}(\mathcal{C})$, the mapping

$$\mathrm{Hom}(C, B) \to \mathrm{Hom}(C, A_1) \times \mathrm{Hom}(C, A_2), \ h \mapsto (\pi_1 \circ h, \pi_2 \circ h)$$

is a bijection. It is injective, because if $h, h' \colon C \to B$ are such that $\pi_i \circ h = \pi_i \circ h'$ for $i = 1, 2$, then $h = 1_B \circ h = (\varepsilon_1 \circ \pi_1 + \varepsilon_2 \circ \pi_2) \circ h = \varepsilon_1 \circ \pi_1 \circ h' + \varepsilon_2 \circ \pi_2 \circ h' = h'$. It is surjective, because if $f \colon C \to A_1$ and $g \colon C \to A_2$, then $\pi_1 \circ (\varepsilon_1 \circ f + \varepsilon_2 \circ g) = f$ and $\pi_2 \circ (\varepsilon_1 \circ f + \varepsilon_2 \circ g) = g$.

(ii) Let (B, π_1, π_2) be a product of A_1 and A_2. By the universal property there exists a unique morphism $\varepsilon_1 \colon A_1 \to B$ such that $\pi_1 \circ \varepsilon_1 = 1_{A_1}$ and $\pi_2 \circ \varepsilon_1 = 0_{A_1, A_2}$. Similarly, there exists a unique morphism $\varepsilon_2 \colon A_2 \to B$ such that $\pi_2 \circ \varepsilon_2 = 1_{A_2}$, $\pi_1 \circ \varepsilon_2 = 0_{A_2, A_1}$. Then $(B, \pi_1, \pi_2, \varepsilon_1, \varepsilon_2)$ is a biproduct of A_1 and A_2, because $\varepsilon_1 \circ \pi_1 + \varepsilon_2 \circ \pi_2 = 1_B$: they are both $B \to B$, and they are equal if and only if $\pi_1 \circ (\varepsilon_1 \circ \pi_1 + \varepsilon_2 \circ \pi_2) = \pi_1 \circ 1_B$, if and only if $1_{A_2} \circ \pi_1 + 0 \circ \pi_2 = \pi_1 = \pi_1 \circ 1_B$, which is true.

□

Passing to the dual category, one sees that $(B, \pi_1, \pi_2, \varepsilon_1, \varepsilon_2)$ is a biproduct of A_1 and A_2 in $\mathcal{C}$ if and only if $(B, \varepsilon_1, \varepsilon_2, \pi_1, \pi_2)$ is a biproduct of A_1 and A_2 in $\mathcal{C}^{\mathrm{op}}$. Moreover, applying Lemma 5.3.11 to the dual categories, one finds that if $(B, \pi_1, \pi_2, \varepsilon_1, \varepsilon_2)$ is a biproduct of A_1 and A_2 in $\mathcal{C}$, then $(B, \varepsilon_1, \varepsilon_2)$ is a coproduct of A_1, A_2, and every coproduct $(B, \varepsilon_1, \varepsilon_2)$ of A_1, A_2 in $\mathcal{C}$ can be completed to a biproduct $(B, \pi_1, \pi_2, \varepsilon_1, \varepsilon_2)$.

Example 5.3.12

We will now give an example of two preadditive categories $\mathcal{C}$ and $\mathcal{D}$ that are isomorphic categories, but that are not isomorphic as preadditive categories. The categories $\mathcal{C}$ and $\mathcal{D}$ will have both a unique object $*$, so that any functor $F \colon \mathcal{C} \to \mathcal{D}$ will send the object $*$ of $\mathcal{C}$ to the object $*$ of $\mathcal{D}$. It follows that the functor $F \colon \mathcal{C} \to \mathcal{D}$ is completely determined by the mapping $F_{**} \colon \mathrm{Hom}_{\mathcal{C}}(*, *) \to \mathrm{Hom}_{\mathcal{D}}(*, *)$. Thus the categories $\mathcal{C}$ and $\mathcal{D}$ are completely determined by the sets $\mathrm{Hom}_{\mathcal{C}}(*, *)$ and $\mathrm{Hom}_{\mathcal{D}}(*, *)$, which must be rings if we want $\mathcal{C}$ and $\mathcal{D}$ to be preadditive categories. Then F is a functor if and only if the mapping $F_{**} \colon \mathrm{Hom}_{\mathcal{C}}(*, *) \to \mathrm{Hom}_{\mathcal{D}}(*, *)$ is a monoid morphism with respect to multiplication, F is a category isomorphism if and only if the mapping $F_{**} \colon \mathrm{Hom}_{\mathcal{C}}(*, *) \to \mathrm{Hom}_{\mathcal{D}}(*, *)$ is a monoid isomorphism with respect to multiplication, F is an additive functor if and only if F_{**} is a ring morphism, and F is an isomorphism of preadditive categories if and only if the mapping F_{**} is a ring isomorphism. Therefore, in order to give an example of two preadditive categories $\mathcal{C}$ and $\mathcal{D}$ that are isomorphic categories, but that are not isomorphic as preadditive categories, it suffices to give and example of two rings R and S that are not isomorphic as rings, but whose multiplicative monoids are isomorphic.

For instance, let k be a countable field and $R := k[x]$, $S := k[x, y]$ the rings of polynomials in one and two indeterminates, respectively. The rings R and S are both countable unique factorization domains with infinitely many pair-wise non-associate polynomials. More precisely, in R there exists a countable set P_R of polynomials such that every non-zero element of R can be written as a product of a non-zero element of k and finitely many elements in P_R, in a unique way. This means that the multiplicative monoid of R is isomorphic to $\{0\} \cup (k^* \times \mathbb{N}^{(\mathbb{N})})$. Similarly for S:

the multiplicative monoid of S is isomorphic to $\{0\} \cup (k^* \times \mathbb{N}^{(\mathbb{N})})$. Thus R and S are isomorphic multiplicative monoids, but they are not isomorphic as rings, because, for instance, R is a principal ideal domain while S has ideals that are not principal, for instance the maximal ideal generated by x and y.

Hence, if $\mathcal{C}$ and $\mathcal{D}$ are the preadditive categories with a unique object $*$, $\mathrm{Hom}_\mathcal{C}(*,*) := R$ and $\mathrm{Hom}_\mathcal{D}(*,*) := S$, and in which morphism composition is the ring multiplication in the two rings R and S, then $\mathcal{C}$ and $\mathcal{D}$ are isomorphic categories, but are not isomorphic as preadditive categories. In other words, $\mathcal{C}$ is a preadditive in which the ring structure on $\mathrm{Hom}_\mathcal{C}(*,*)$ can be given in different ways. That is, the abelian group structure on $\mathrm{Hom}_\mathcal{C}(*,*)$ is not uniquely determined by the category $\mathcal{C}$.

5.4 Additive Categories

By Lemma 5.3.11, two objects A_1 and A_2 of a preadditive category have a biproduct

$$A_1 \underset{\varepsilon_1}{\overset{\pi_1}{\rightleftarrows}} B \underset{\varepsilon_2}{\overset{\pi_2}{\rightleftarrows}} A_2$$

if and only if they have a product

$$A_1 \xleftarrow{\pi_1} B \xrightarrow{\pi_2} A_2,$$

if and only if they have a coproduct

$$A_1 \xrightarrow{\varepsilon_1} B \xleftarrow{\varepsilon_2} A_2.$$

In this case, we denote by $(A_1 \oplus A_2, \pi_1, \pi_2, \varepsilon_1, \varepsilon_2)$ such a biproduct. Sometimes we will omit the maps and we will simply write $A_1 \oplus A_2$ for such a 5-tuple.

A category $\mathcal{C}$ is an *additive category* if it is preadditive, has a null object, and every pair (A, B) of objects of $\mathcal{C}$ has a coproduct in $\mathcal{C}$. Equivalently, if it is preadditive, has a null object, and every pair of objects of $\mathcal{C}$ has a product in $\mathcal{C}$. We will write $A \oplus B$ for $A \prod B \cong A \coprod B$. For instance, the category Mod-R and its full subcategory whose objects are all free modules are both additive categories.

We will now prove that the abelian group structure on the abelian groups $\mathrm{Hom}_\mathcal{C}(A_1, A_2)$ of an additive category $\mathcal{C}$ is uniquely determined. This is not true for preadditive categories, as we have seen in Example 5.3.12.

Lemma 5.4.1

If $\mathcal{C}$ is an additive category and $A, B \in Ob(\mathcal{C})$, then the abelian group structure on the sets $\mathrm{Hom}_\mathcal{C}(A, B)$ is uniquely determined.

5.4 Additive Categories

Proof Let $(A \oplus A, \pi_1, \pi_2, \varepsilon_1, \varepsilon_2)$ and $(B \oplus B, \pi'_1, \pi'_2, \varepsilon'_1, \varepsilon'_2)$ be the biproducts of A and A, and B and B, respectively. We claim that if $f, g \in \text{Hom}_{\mathcal{C}}(A, B)$, then $f + g$ can be written as the composite morphism of the following three morphisms:

(1) the morphism $\Delta_A \colon A \to A \oplus A$, where Δ_A is unique morphism such that $\pi_1 \circ \Delta_A = 1_A$ and $\pi_2 \circ \Delta_A = 1_A$;
(2) the morphism $s \colon A \oplus A \to B \oplus B$ defined by $s = \varepsilon'_1 \circ f \circ \pi_1 + \varepsilon'_2 \circ g \circ \pi_2$;
(3) the morphism $\nabla_B \colon B \oplus B \to B$, where ∇_B is the unique morphism such that $\nabla_B \circ \varepsilon'_1 = 1_B$ and $\nabla_B \circ \varepsilon'_2 = 1_B$.

In order to prove the claim it suffices to notice that $\nabla_B \circ s \circ \Delta_A = \nabla_B \circ (\varepsilon'_1 \circ f \circ \pi_1 + \varepsilon'_2 \circ g \circ \pi_2) \circ \Delta_A = \nabla_B \circ \varepsilon'_1 \circ f \circ \pi_1 \circ \Delta_A + \nabla_B \circ \varepsilon'_2 \circ g \circ \pi_2 \circ \Delta_A = f + g$.

This proves the claim. Now s can be defined also without using the addition of $\text{Hom}_{\mathcal{C}}(A, B)$, as follows. By the universal property of product, there is a unique morphism $s' \colon A \to B \oplus B$ such that $\pi'_1 s' = f$ and $\pi'_2 s' = 0$. Similarly, there is a unique morphism $s'' \colon A \to B \oplus B$ such that $\pi'_1 s'' = 0$ and $\pi'_2 s'' = g$. By the universal property of coproduct, there is a unique morphism $s \colon A \oplus A \to B \oplus B$ such that $s\varepsilon_1 = s'$ and $s\varepsilon_2 = s''$. Looking at the proof of Lemma 5.3.11(i) one sees that $s' = \varepsilon'_1 f$. Similarly, $s'' = \varepsilon'_2 g$ and $s = s'\pi_1 + s''\pi_2$. Thus $s = \varepsilon'_1 f \pi_1 + \varepsilon'_1 g \pi_2$, as desired. Therefore $f + g = \nabla_B s \Delta_A$ is determined without using the addition of $\text{Hom}_{\mathcal{C}}(A, B)$. □

As a corollary to this lemma, we get that if, instead of the biproducts $A \oplus A$, $B \oplus B$, we choose other biproducts $(A \oplus A)'$ and $(B \oplus B)'$, so that for instance we have the corresponding morphism $(\Delta_A)' \colon A \to (A \oplus A)'$, then $\nabla_B \circ (f \oplus g) \circ \Delta_A = (\nabla_B)' \circ (f \oplus g)' \circ (\Delta_A)'$. Thus the three morphisms ∇_B, $f \oplus g$ and Δ_A change, but their composite morphism does not.

We now consider the behaviour of biproducts with respect to additive functors. Recall that a functor $F \colon \mathcal{C} \to \mathcal{D}$ between two preadditive categories is additive if and only if, for every pair of objects A, B of $\mathcal{C}$, the mapping $F_{A,B} \colon \text{Hom}_{\mathcal{C}}(A, B) \to \text{Hom}_{\mathcal{D}}(FA, FB)$, induced by F, is an abelian group morphism.

> **Lemma 5.4.2**
> Let $\mathcal{C}$ and $\mathcal{D}$ be two additive categories. A functor $F \colon \mathcal{C} \to \mathcal{D}$ is additive if and only if it preserves biproducts, i.e., for any two objects A and B of $\mathcal{C}$, if $(A \oplus B, \pi_A, \pi_B, \varepsilon_A, \varepsilon_B)$ is a biproduct of A and B, then $(F(A \oplus B), F(\pi_A), F(\pi_B), F(\varepsilon_A), F(\varepsilon_B))$ is a biproduct of $F(A)$ and $F(B)$.

Proof Suppose first F additive. Without lost of generality, assume F covariant. Consider a biproduct $(A \oplus B, \pi_A, \pi_B, \varepsilon_A, \varepsilon_B)$ of A and B. By definition, $\pi_A \circ \varepsilon_A = 1_A$, $\pi_B \circ \varepsilon_B = 1_B$ and $\varepsilon_A \circ \pi_A + \varepsilon_B \circ \pi_B = 1_{A \oplus B}$. Since F is covariant and additive, we get that $F\pi_A \circ F\varepsilon_A = 1_{FA}$, $F\pi_B \circ F\varepsilon_B = 1_{FB}$ and $F\varepsilon_A \circ F\pi_A + F\varepsilon_B \circ F\pi_B =$

$1_{F(A \oplus B)}$. Thus $(F(A \oplus B), F(\pi_A), F(\pi_B), F(\varepsilon_A), F(\varepsilon_B))$ is a biproduct of $F(A)$ and $F(B)$.

Conversely, let $f, g: A \to B$. As we have seen in the proof of Lemma 5.4.1, we have that $f + g = \nabla_B \circ (f \oplus g) \circ \Delta_A$ and $F(f) + F(g) = \nabla_{F(B)} \circ (F(f) \oplus F(g)) \circ \Delta_{F(A)}$. Hence it suffices to prove that $F \nabla_B = \nabla_{F(B)}$, $F(f \oplus g) = F(f) \oplus F(g)$ and $F(\Delta_A) = \Delta_{F(A)}$. This follows from the fact that $(F(A \oplus B), F(\pi_A), F(\pi_B), F(\varepsilon_A), F(\varepsilon_B))$ is a biproduct of $F(A)$ and $F(B)$ and by the definition of the written morphisms. □

5.5 Equalizers, Coequalizers, Kernels and Cokernels

In this section, we introduce the notion of kernel and cokernel of a morphism in a category.

Definition 5.5.1

Let $\mathcal{C}$ be a category, $A, B \in \mathrm{Ob}(\mathcal{C})$ and let $f, g: A \to B$ be two morphisms in $\mathcal{C}$. The *equalizer* of f and g is a morphism $e: C \to A$ for some object C such that $f \circ e = g \circ e$, satisfying the following universal property: for every object D and every morphism $t: D \to A$ such that $f \circ t = g \circ t$, there exists a unique morphism $m: D \to C$ such that $t = e \circ m$.

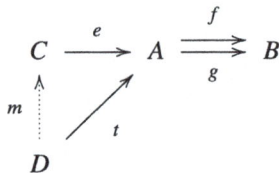

If an equalizer of two morphisms f and g exists, it is unique up to isomorphism in the following sense: if $e: C \to A$ and $e': C' \to A$ are both equalizers of f and g, then there exists a unique isomorphism $\varphi: C \to C'$ such that $\varphi \circ e = e'$.

Examples 5.5.2
(a) In the category $\mathcal{C} = \mathbf{Set}$ of sets, the equalizer of two mappings $f, g: X \to Y$ exists always. To see it, consider the set Z of all $x \in X$ such that $f(x) = g(x)$, and take the inclusion of Z into X as the equalizer $e: Z \to X$. If $l: W \to X$ is such that $g \circ l = f \circ l$, then $g(l(w)) = h(l(w))$ for every $w \in W$, so that $l(w) \in Z$. Thus $l(W) \subseteq Z$. Now let $m: W \to Z$ be the mapping obtained from $l: W \to X$ restricting the codomain to Z.
(b) In the category $\mathcal{C} = \mathbf{Top}$ of topological spaces, the equalizer of two continuous mappings $f, g: X \to Y$ is in (a), where on Z there is the topology induced by the topology of X (the inclusion mapping e is continuous).
(c) In the category $\mathbf{Ab}$ of abelian groups, the equalizer of two homomorphisms $f, g: G \to H$ is the inclusion $e: \ker(f - g) \hookrightarrow G$.

5.5 Equalizers, Coequalizers, Kernels and Cokernels

We can also consider the following dual notion, the notion of a coequalizer.

Definition 5.5.3

Let $\mathcal{C}$ be any category, and $A, B \in \text{Ob}(\mathcal{C})$. A coequalizer of the morphisms $f, g: A \to B$ is a morphism $h: B \to C$ such that $h \circ f = h \circ g$ and, for any morphism $t: B \to D$ in $\mathcal{C}$ with $t \circ f = t \circ g$, there exists a unique morphism $m: C \to D$ such that $m \circ h = t$.

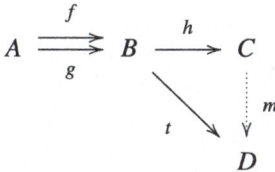

Example 5.5.4
In the category $\mathcal{C} = \textbf{Set}$ of sets, let $f, g: X \to Y$ be two morphisms. In $Y \times Y$, let $\sim$ be the least equivalence relation containing $\{(f(x), g(x)) \mid x \in X\}$. Let $\pi: Y \to Y/\sim$ be the canonical projection onto the quotient set $Y/\sim$. This is a coequalizer of f and g, because $\pi \circ f = \pi \circ g$ if and only if $\pi(f(x)) = \pi(g(x))$ for every $x \in X$. This is the case if and only if $[f(x)]_\sim = [g(x)]_\sim$.

In the category $\mathcal{C} = \textbf{Top}$ of topological spaces, the coequalizer of two morphisms f and g is the same as for **Set** with the quotient topology on $Y/\sim$.

In the category $\mathcal{C} = \textbf{Ab}$ of abelian groups, let $f, g: G \to H$ be two morphisms. The coequalizer of f and g is the canonical projection $\pi: H \to H/(f-g)(G)$.

Definition 5.5.5

Let $\mathcal{C}$ be a category with a null object Z. The *kernel* of a morphism $f: A \to B$ is the equalizer of f and the zero morphism $\zeta_{A,B}: A \to B$. The *cokernel* of f is the coequalizer of f and $\zeta_{A,B}$.

Thus the kernel of $f: A \to B$ is any morphism $k: K \to A$ such that (1) $fk = \zeta_{K,B}$, and (2) for every morphism $t: D \to A$ such that $ft = \zeta_{D,B}$ there exists a unique morphism $m: D \to K$ with $t = km$.

When a kernel of a morphism exists, it is unique up to isomorphism.

Example 5.5.6
Let $\textbf{Set}_*$ be the category whose objects are all pairs (X, x_0) where X is a set and $x_0 \in X$, and whose morphisms $f: (X, x_0) \to (Y, y_0)$ are all morphisms $f: X \to Y$ in **Set** (that is, all mappings $f: X \to Y$) such that $f(x_0) = y_0$, with the usual composition. Null objects turn out to be the pairs $(\{x_0\}, x_0)$. We will now determine the kernel of $f: (X, x_0) \to (Y, y_0)$. The zero morphism $(X, x_0) \to (Y, y_0)$ is the mapping $\zeta_{X,Y}: X \to Y$ that sends all elements of X to y_0. Set $X' := \{x \in X \mid f(x) = y_0\} \subseteq X$, so that $x_0 \in X'$. It is now easy to check that the inclusion $(X', x_0) \overset{h}{\hookrightarrow} (X, x_0)$ is the kernel of $f: (X, x_0) \to (Y, y_0)$. For instance, the composition $f \circ h = f|_{X'}$ is the constant mapping y_0, that is, the zero mapping $\zeta_{X',Y}$. Let us determine the cokernel of $f: (X, x_0) \to (Y, y_0)$. Let $\mathcal{F}$ be the partition of Y having a block equal to $f(X) \subseteq Y$ and all the other blocks of cardinality 1. Equivalently, $\mathcal{F}$ is the quotient set $Y/\sim$, where $\sim$ is the equivalence relation on Y defined, for every $y, y' \in Y$, by $y \sim y'$ if either $y = y'$, or $y \in f(X)$ and

$y' \in f(X)$. Consider the object $(Y/\sim, f(X))$ of **Set**$_*$. Then the canonical mapping $Y \to Y/\sim$ turns out to be the cokernel of the morphism f.

Exercise 5.5.7
Show that in a preadditive category $\mathcal{C}$ with a null object, every equalizer is a kernel and every coequalizer is a cokernel. More precisely, every Hom-set is an additive abelian group, and hence if h is an equalizer of f and g, then $f \circ h = g \circ h$ if and only if $(f - g) \circ h = 0$. Thus the equalizer of f and g is the kernel of $f - g$.

Exercise 5.5.8
Show that every equalizer is a monomorphism and every coequalizer is an epimorphism. In particular, every kernel is a monomorphism and every cokernel is an epimorphism.

Exercise 5.5.9
Let $\mathcal{C}$ be a category with a null object and in which every pair of morphisms $f, g\colon X \to Y$ has an equalizer and a coequalizer. Show that the equalizer of the morphisms f and g in the dual category $\mathcal{C}^{\mathrm{op}}$ is the coequalizer of f and g in the category $\mathcal{C}$. Similarly, the coequalizer of f and g in the dual category $\mathcal{C}^{\mathrm{op}}$ is the equalizer of f and g in the category $\mathcal{C}$. In particular, for every morphism u of $\mathcal{C}$, the kernel of u in $\mathcal{C}^{\mathrm{op}}$ is the cokernel of u in $\mathcal{C}$, and the cokernel of u in $\mathcal{C}^{\mathrm{op}}$ is the kernel of u in $\mathcal{C}$.

Exercise 5.5.10
Let $\mathcal{C}$ be a preadditive category with a null object $0_\mathcal{C}$ and let $f\colon A \to B$ be a morphism in $\mathcal{C}$. Then f is a monomorphism if and only if $\zeta_{0,A}\colon 0_\mathcal{C} \to A$ is a kernel for f.

Exercise 5.5.11
Let $\mathcal{C}$ be a preadditive category with a null object $0_\mathcal{C}$ and let $f\colon A \to B$ be a morphism in $\mathcal{C}$ that is the kernel of a morphism $g\colon B \to C$. Assume that coker f exists. Then $f = \ker(\mathrm{coker}\, f)$.

5.6 Abelian Categories

Definition 5.6.1

A *preabelian* category is an additive category $\mathcal{C}$ in which every morphism has a kernel and a cokernel.

If $\mathcal{C}$ is a preabelian category and $f\colon A \to B$ is a morphism, then $\mathrm{coker}(f)$ has a kernel. Define the *image* of f as

$$\mathrm{im}(f) := \ker(\mathrm{coker}(f)).$$

Dually

$$\mathrm{coim}(f) := \mathrm{coker}(\ker(f))$$

is the *coimage* of f. Notice that $\mathrm{im}(f)$ is a subobject of B. Hence there are morphisms $p\colon A \to \mathrm{coim}(f)$ and $k\colon \mathrm{im}(f) \to B$.

5.6 Abelian Categories

Example 5.6.2
If C is the category **Ab** of abelian groups and $f: G \to H$ is a morphism, then the embedding $\ker(f) \hookrightarrow G$ is a kernel of f, and the canonical projection $H \to H/f(G)$ is a cokernel of f. Hence **Ab** is a preabelian category. The inclusion $k: f(G) \hookrightarrow H$ is an image of f, and the canonical projection $p: G \to G/\ker(f)$ is a coimage of f.

> **Proposition 5.6.3** (Canonical Factorization of Morphisms)
> Let C be a preabelian category. For any morphism $f: A \to B$ there exists a unique morphism $\overline{f}: \mathrm{coim}(f) \to \mathrm{im}(f)$ such that $f = k \circ \overline{f} \circ p$.

Proof Uniqueness. If $f': \mathrm{coim}(f) \to \mathrm{im}(f)$ also has the property that $f = k \circ f' \circ p$, then $k \circ \overline{f} \circ p = k \circ f' \circ p$. But p is an epimorphism, so $k \circ \overline{f} = k \circ f'$, and k is a monomorphism, hence $f' = \overline{f}$.

Existence. Consider the following commutative diagram:

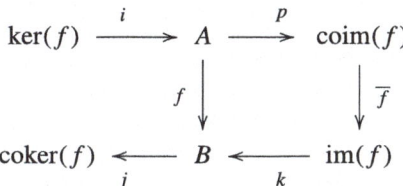

Then $f \circ i = 0$. Since $p = \mathrm{coker}(i)$, f factors through the cokernel of i, so $f = f' \circ p$. Since $j \circ f = 0$, we have $j \circ f' \circ p = 0$, so $j \circ f' = 0$. Thus f' factors uniquely through the kernel of j, i.e., $f' = k \circ \overline{f}$. Then $f = f' \circ p = k \circ \overline{f} \circ p$. □

Example 5.6.4
Let $C = \mathbf{Ab}$. If $f: A \to B$ is an abelian group morphism, $\overline{f}: A/\ker(f) \to f(A)$ is the induced isomorphism, which sends $a + \ker(f)$ to $f(a)$.

The unique morphism $\overline{f}: \mathrm{coim}(f) \to \mathrm{im}(f)$ in Proposition 5.6.3 is called the *parallel* of f. An *abelian* category is a preabelian category in which the parallels of all morphisms are isomorphisms. Thus a category C is abelian if:

(1) C is preadditive.
(2) Every finite family of objects of C has a product. Equivalently, C has a null object and any two objects of C have a product.
(3) Every morphism has a kernel and a cokernel. This implies that if $f: A \to B$ is a morphism there exists a unique morphism $\overline{f}: \mathrm{coim}(f) \to \mathrm{im}(f)$ such that $f = i \circ \overline{f} \circ p$ where $p: A \to \mathrm{coim}(f)$, $i: \mathrm{im}(f) \to B$ are the coimage and the image of f respectively.
(4) $\overline{f}$ is an isomorphism for any morphism f.

Notice that we can replace Condition (4) in the previous definition with one of the following:
(4′) Every morphism $f: A \to B$ can be written as $f = g \circ h$ where g is the kernel of some morphism and h is the cokernel of some morphism.
(4″) Every monomorphism is a kernel and every epimorphism is a cokernel.

Definition 5.6.5

A sequence

$$A \xrightarrow{f} B \xrightarrow{g} C$$

in $\mathcal{C}$ is *exact* if for any morphism $h: D \to B$, h is an image of f if and only if h is a kernel of g.

If $F: \mathcal{C} \to \mathcal{C}'$ is an additive functor, where $\mathcal{C}$ and $\mathcal{C}'$ are abelian, F is an *exact functor* if it sends exact sequences into exact sequences.

Exercises on Abelian Categories

Let R be a commutative integral domain. For any R-module M_R, the *torsion submodule* $t(M_R)$ of M_R is consists of all elements $x \in M_R$ for which there exists a non-zero element $r \in R$ such that $rx = 0$ (Check that it is indeed a submodule of M_R!) A module M_R is *torsion* if $t(M_R) = M_R$. A module M_R is *torsion-free* if $t(M_R) = 0$, that is, if for every $r \in R$ and $x \in M_R$, $rx = 0$ implies $r = 0$ or $x = 0$. Equivalently, a module M_R is torsion-free if for every non-zero element $r \in R$, multiplication by r is an injective homomorphism $\lambda_r: M_R \to M_R$. For instance, every free module over a commutative integral domain is torsion-free. For every module M_R over an arbitrary integral domain R, the submodule $t(M_R)$ is a torsion module, and the quotient $M_R/t(M_R)$ is torsion-free. If R is a commutative integral domain, then, for every R-module morphism $f: M_R \to N_R$ one has that $f(t(M_R)) \subseteq t(N_R)$. Hence there is a functor Mod-$R \to$ Mod-R that associate to every R-module M_R its torsion submodule $t(M_R)$ and to every morphism f in Mod-R the restriction of f to the torsion submodules.

Exercise 5.6.6
Let $\mathcal{F}$ be the full subcategory of **Ab** whose objects are all free abelian groups of finite rank.
(a) Show that $\mathcal{F}$ is an additive category.
(b) Show that a morphism $f: G \to H$ in the category $\mathcal{F}$ is a monomorphism in $\mathcal{F}$ if and only if it is a monomorphism in **Ab**, that is, it is an injective mapping.
(c) Show that every morphism $f: G \to H$ in the category $\mathcal{F}$ has a kernel in $\mathcal{F}$, which coincides with the kernel in **Ab**. A morphism $g: L \to G$ in $\mathcal{F}$ is a kernel in $\mathcal{F}$ if and only if it is an injective mapping and $H/f(G)$ is torsion-free.
(d) Show that a morphism $f: G \to H$ in the category $\mathcal{F}$ is an epimorphism in $\mathcal{F}$ if and only if the group $H/f(G)$ is a torsion group.

5.6 Abelian Categories

(e) Show that every morphism $f: G \to H$ in the category $\mathcal{F}$ has a cokernel in $\mathcal{F}$, which coincides with the canonical projection

$$\pi: H \to (H/f(G))/t(H/f(G)).$$

Hence a morphism $k: K \to G$ in the category $\mathcal{F}$ is a cokernel in $\mathcal{F}$ if and only if it is a surjective mapping.

(f) Show that multiplication by 2 is an epimorphism $\mathbb{Z} \to \mathbb{Z}$ in $\mathcal{F}$ that is not a cokernel. In particular, $\mathcal{F}$ is a preabelian category that is not an abelian category.

Exercise 5.6.7

A *topological abelian group* is a triple $(G, +, \tau)$, where $(G, +)$ is an abelian group, (G, τ) is a topological space and the sum $G \times G \to G$, $(g, g') \mapsto g + g'$, and the opposite $G \to G, g \mapsto -g$, are continous mappings. Here $G \times G$ is endowed with the product topology. Let $\mathcal{C}$ be the category of all Hausdorff topological abelian groups, i.e., the category whose objects are all Hausdorff topological abelian groups and in which the morphisms are the *continuous homomorphisms*, that is, the mappings $f: G \to G'$ that are both continous mappings and group homomorphisms. The composition is the usual compositions of two mappings. It is easily seen that $\mathcal{C}$ is a category.

Show that:

(a) The sum of two continuous group morphisms is a continous group morphism. The category $\mathcal{C}$ is preadditive. [*Hint.* The sum $f + g$ of two continuous morphisms $f, f': G \to H$ is the composite mapping of the diagonal mapping $\Delta: G \to G \times G$, $\Delta(x) = (x, x)$, the product mapping $f \times f': G \times G \to H \times H$, $(f \times f')(x, y) = (f(x), f(y))$, and the sum $H \times H \to H$. Show that these three mappings are continuous.]

(b) The category $\mathcal{C}$ is additive (determine the biproducts!). The kernel of a morphism $f: G \to H$ in the category $\mathcal{C}$ is the inclusion $\ker(f) \hookrightarrow G$, where $\ker(f)$ is the kernel of f (in the sense of Group Theory) endowed with the topology induced from the topology of G. Hence the kernels in $\mathcal{C}$ are the topological embeddings, that is, the continuous homomorphism $f: G \to H$ that induce a homeomorphism between G and $f(G)$. Here $f(G)$ is endowed with the topology induced from the topology of H. Notice that every morphism in $\mathcal{C}$ has a kernel in $\mathcal{C}$.

(c) The monomorphisms in $\mathcal{C}$ are the continuous homomorphims that are injective mappings.

(d) The identity $\iota: \mathbb{R} \to \mathbb{R}$ is a monomorphism of the topological group $\mathbb{R}$ of real numbers with the discrete topology into the topological group $\mathbb{R}$ of real numbers with the usual topology. This monomorphism is not a kernel in $\mathcal{C}$. Hence the category $\mathcal{C}$ is not abelian.

(e) The cokernel of a morphism $f: G \to H$ in $\mathcal{C}$ is the canonical projection $\pi: H \to H/f(G)$, where $H/f(G)$ is endowed with the quotient topology, that is, the finest topology that makes π continuous (a subset U of $H/f(G)$ is an open set if and only if its inverse image $\pi^{-1}(U)$ is an open subset of H). Hence the cokernels in $\mathcal{C}$ are the onto continuous homomorphisms $f: G \to H$ with the property that a subset U of H is an open subset of H if and only if its inverse image $f^{-1}(U)$ is an open subset of G. Every morphism in $\mathcal{C}$ has a cokernel in $\mathcal{C}$.

(f) The epimorphisms in $\mathcal{C}$ are the continuous homomorphims that are onto mappings.

(g) The identity ι in (d) is an epimorphism, but not a cokernel in $\mathcal{C}$. Thus that mapping ι is a monomorphism and an epimorphism, but not a kernel, not a cokernel, and not an isomorphism.

(h) The category $\mathcal{C}$ of topological groups is preabelian, but not abelian.

Exercise 5.6.8

Let I be a directed set, R a ring and $(\text{Mod-}R)^I$ the category of all direct systems of right R-modules over the directed set I.

(1) Show that the category $(\text{Mod-}R)^I$ is abelian.
(2) Show that a sequence $\mathbb{M} \xrightarrow{\alpha} \mathbb{N} \xrightarrow{\beta} \mathbb{P}$ is exact in $(\text{Mod-}R)^I$ if and only if all the sequences $M_i \longrightarrow N_i \longrightarrow P_i, i \in I$, are exact in Mod-$R$.

(3) Show that the additive functor

$$\varinjlim \colon (\text{Mod-}R)^I \to \text{Mod-}R$$

of Proposition 4.6.9 is exact.

Sketch of Solution (1) In the category $(\text{Mod-}R)^I$, the objects are the direct systems $\mathbb{M} := (M_i, \mu_{ij})$. If $\mathbb{M} := (M_i, \mu_{ij})$ and $\mathbb{N} = (N_i, \nu_{ij})$ are two direct systems of right R-modules over the directed set I, a morphism $\Phi \colon \mathbb{M} \to \mathbb{N}$ is by definition a family $\Phi = (\varphi_i)_{i \in I}$ of R-module morphisms $\varphi_i \colon M_i \to N_i$ such that $\nu_{ij} \circ \varphi_i = \varphi_j \circ \mu_{ij}$ for every $i \leq j$. Thus $\text{Hom}_{(\text{Mod-}R)^I}(\mathbb{M}, \mathbb{N})$ is a subgroup of $\prod_{i \in I} \text{Hom}_{\text{Mod-}R}(M_i, N_i)$. The composition is defined, for $\Phi = (\varphi_i)_{i \in I} \colon \mathbb{M} \to \mathbb{N}$ and $\Psi = (\psi_i)_{i \in I} \colon \mathbb{N} \to \mathbb{P}$, by $\Psi \circ \Phi := (\psi \circ \varphi_i)_{i \in I}$. The identity of an object $\mathbb{M}$ is $1_{\mathbb{M}} = (1_{M_i} \colon M_i \to M_i)_{i \in I}$. The null object is $\mathbb{O} = (0_i, \mu_{ij} = 0)$. The product, coproduct, biproduct of $\mathbb{M}$ and $\mathbb{N}$ is $\mathbb{M} \prod \mathbb{N} = (M_i \oplus N_i, \mu_{ij} \oplus \nu_{ij} \colon M_i \oplus N_i \longrightarrow M_j \oplus N_j)$. The kernel of $\Phi \colon \mathbb{M} \to \mathbb{N}$ is the inclusion k of

$$\mathbb{K} = (\ker(\varphi_i), \mu_{ij}|_{\ker(\varphi_i)} \colon \ker(\varphi_i) \to \ker(\varphi_j))_{i \in I} \hookrightarrow \mathbb{M} = (M_i, \mu_{ij})$$

into $\mathbb{M}$. (Notice that $\mu_{ij}(\ker \varphi_i) \subseteq \ker \varphi_j$, because if $x \in \ker \varphi_i$, then $\varphi_j \mu_{ij}(x) = \nu_{ij} \varphi_i(x) = 0$. Here $k = (k_i \colon \ker \varphi_i \hookrightarrow M_i)$. Similarly for cokernels. We leave the details to the reader.

(3) Let $\mathbb{M} \xrightarrow{\alpha} \mathbb{N} \xrightarrow{\beta} \mathbb{P}$ be an exact sequence in $(\text{Mod-}R)^I$. Apply the direct limit and get the sequence

$$M \xrightarrow{\varphi} N \xrightarrow{\psi} P,$$

where M, N, P is the direct limits of $\mathbb{M}, \mathbb{N}, \mathbb{P}$ respectively. In order to show that $\ker(\psi) \subseteq \text{im}(\varphi)$, fix an element $y \in \ker(\psi)$. There are $i \in I$ and $y_i \in N_i$ such that $y = \nu_i(y_i)$, so that $\pi_i(\psi_i(y_i)) = \psi(\nu_i(y_i)) = 0 \in P$. Thus there exists $j \geq i$ with $\pi_{ij}(\psi_i(y_i)) = 0$, i.e., $\psi_j(\nu_{ij}(y_i)) = 0$. Hence $\nu_{ij}(y_i) \in \ker(\psi_j) = \text{im}(\varphi_j)$. This shows that $\nu_{ij}(y_i) = \varphi_j(x_j)$ for a suitable $x_j \in M_j$. Then

$$y = \nu_i(y_i) = \nu_j(\nu_{ij}(y_i)) = \nu_j(\varphi_j(x_j)) = \varphi(\mu_j(x_j)).$$

It follows that $y \in \text{im}(\varphi)$, as we wanted to prove.

Exercise 5.6.9
Let $\mathcal{C}$ be a preadditive category with a null object. Let $f \colon A \to B$ a fixed morphism in $\mathcal{C}$. Let $\mathcal{D}$ be the category defined by
(1) $\text{Ob}(\mathcal{D}) = \{(C, g) \mid C \in \text{Ob}(\mathcal{C}), g \in \text{Hom}_\mathcal{C}(C, A), fg = 0\}$;
(2) $\text{Hom}_\mathcal{D}((C, g), (C', g')) = \{h \mid h \in \text{Hom}_\mathcal{C}(C, C'), g = g'h\}$;
(3) the composition in $\mathcal{D}$ is induced by the composition in $\mathcal{C}$.
  (a) Show that $\mathcal{C}$ is a category.
  (b) Show that (C', g') is a terminal object in $\mathcal{D}$ if and only if $g' \colon C' \to A$ is a kernel for f in $\mathcal{C}$.

5.6 Abelian Categories

Exercise 5.6.10
Let $\mathcal{C}$ be an abelian category. Show that a sequence

$$0 \to A \xrightarrow{f} B \xrightarrow{g} C \tag{5.1}$$

in $\mathcal{C}$ is exact if and only if the sequence

$$0 \to \operatorname{Hom}_{\mathcal{C}}(X, A) \xrightarrow{\operatorname{Hom}_{\mathcal{C}}(X, f)} \operatorname{Hom}_{\mathcal{C}}(X, B) \xrightarrow{\operatorname{Hom}_{\mathcal{C}}(X, g)} \operatorname{Hom}_{\mathcal{C}}(X, C) \tag{5.2}$$

is exact in **Ab** for every $X \in \operatorname{Ob}(\mathcal{C})$.

Solution
($\Rightarrow$) Assume that (5.1) is an exact sequence in $\mathcal{C}$.

Step 1. $\operatorname{Hom}_{\mathcal{C}}(X, f)$ *is a monomorphism.* Let $h, h' \colon X \to A$ be two morphisms with $\operatorname{Hom}_{\mathcal{C}}(X, f)(h) = \operatorname{Hom}_{\mathcal{C}}(X, f)(h')$. Then $f \circ h = f \circ g$. Now (5.1) is exact in $\mathcal{C}$, so that $\ker f = 0$, i.e., f is a monomorphism. Thus $h = g$.

Step 2. $\operatorname{im}(\operatorname{Hom}_{\mathcal{C}}(X, f)) \subseteq \ker(\operatorname{Hom}_{\mathcal{C}}(X, g))$. Since the sequence (5.1) is exact, we know that $\operatorname{im}(f) = \ker(g)$. Making use of the canonical decomposition of morphisms in preabelian categories, we can write $f = k \circ \overline{f} \circ p$, where $\overline{f} \colon \operatorname{coim}(f) \to \operatorname{im}(f) = \ker(g)$ is the parallel of f. Thus $g \circ f = g \circ k \circ \overline{f} \circ p = 0 \circ \overline{f} \circ p = 0$. Now Step 2 follows immediately from the fact that $\operatorname{Hom}_{\mathcal{C}}(X, f) \circ \operatorname{Hom}_{\mathcal{C}}(X, g) = \operatorname{Hom}_{\mathcal{C}}(X, f \circ g) = \operatorname{Hom}_{\mathcal{C}}(X, 0) = 0$.

Step 3. $\ker(\operatorname{Hom}_{\mathcal{C}}(X, g)) \subseteq \operatorname{im}(\operatorname{Hom}_{\mathcal{C}}(X, f))$. Let α be an element of $\ker(\operatorname{Hom}_{\mathcal{C}}(X, g))$. This means that $\alpha \colon X \to B$ is a morphism in $\mathcal{C}$ such that $g \circ \alpha = 0$. Hence α factors in a unique way through $\ker(g) = \operatorname{im}(f)$. That is, if $\beta \colon \operatorname{im}(f) \to B$ denotes the image of f, there exists a unique $\gamma \colon X \to \operatorname{im}(f)$ such that $\alpha = \beta \circ \gamma$. Now (5.1) exact implies $\ker(f) = 0$. Thus $\operatorname{coim}(f) = \operatorname{coker}(\ker(f)) = 1_A \colon A \to A$. Thus the canonical decomposition of f is $f = \beta \circ \overline{f} \circ 1_A$. Thus $\alpha = \beta \circ \gamma = \beta \circ \overline{f} \circ \overline{f}^{-1} \circ \gamma = \beta \circ \overline{f} \circ 1_A \circ \overline{f}^{-1} \circ \gamma = f \circ \overline{f}^{-1} \circ \gamma \in \operatorname{im}(\operatorname{Hom}_{\mathcal{C}}(X, f))$.

($\Leftarrow$) Assume that the sequences (5.2) are exact for every $X \in \operatorname{Ob}(\mathcal{C})$. We want to show that (5.1) is exact in $\mathcal{C}$.

Exactness of (5.1) in A We must prove that $\ker(f) = 0$, i.e., that f is a monomorphism. Let $g, h \colon X \to A$ be two morphisms with $f \circ g = f \circ h$. Then $\operatorname{Hom}_{\mathcal{C}}(X, f)(g) = \operatorname{Hom}_{\mathcal{C}}(X, f)(h)$. By hypothesis, $\operatorname{Hom}_{\mathcal{C}}(X, f)$ is a monomorphism. Therefore $g = h$, so f is a monomorphism.

Exactness of (5.1) in B We must prove that $\operatorname{im}(f) = \ker(g)$. Let $c \colon B \to D$ be a cokernel for f, so that $\operatorname{im}(f) = \ker(\operatorname{coker}(f)) = \ker(c)$. Hence we must show that $\ker(g) = \ker(c)$. For this, it suffices to prove that, for every $h \colon X \to B$, one has that $g \circ h = 0$ if and only if $c \circ h = 0$. Now, by hypothesis, the sequence is exact in $\operatorname{Hom}_{\mathcal{C}}(X, B)$, that is, for every $h \colon X \to B$, one has $h \in \ker(\operatorname{Hom}_{\mathcal{C}}(X, g))$ if and only if $h \in \operatorname{im}(\operatorname{Hom}_{\mathcal{C}}(X, f))$. Equivalently, for any $h \colon X \to B$, one has that $g \circ h = 0$ if and only if h factors through f.

Hence we must prove that, for every $h \colon X \to B$, one has that $c \circ h = 0$ if and only if h factors through f.

($\Leftarrow$) If h factors through f, then $h = f \circ l$, so that $c \circ h = c \circ f \circ l = 0$ because c is cokernel of f.

($\Rightarrow$) Assume that $c \circ h = 0$, where c is the cokernel of f. Then h factors through $\ker c$, i.e. h factors through $\ker(\operatorname{coker} f)$, which is f, as is easy to prove. $\square$

Appendices

6.1 Appendix 1: Axiomatic Set Theory

In this first appendix we want to further develop what we have seen on pages 57 and 58 about ZFC and NBG. The axioms of ZFC are the following (To be precise, the axioms are formulas. We have added some comments for clarity):

1. Axiom of extensionality. Two sets are equal if they have the same elements, that is, a set is determined by its elements:

$$\forall x \forall y (\forall z (z \in x \Leftrightarrow z \in y) \Rightarrow x = y).$$

2. Axiom of foundation (or axiom of regularity). Every non-empty set x contains an element y such that x and y are disjoint sets.

$$\forall x [\exists a (a \in x) \Rightarrow \exists y (y \in x \wedge \neg \exists z (z \in y \wedge z \in x))].$$

3. Axiom schema of specification (also called the axiom schema of separation). If z is a set, and ϕ is a property that the elements x of z can have or not have, then there exists a subset y of z containing the x in z which satisfy the property ϕ.

$$\forall z \forall w_1 \ldots w_n \exists y \forall x (x \in y \Leftrightarrow (x \in z \wedge \phi)).$$

Here ϕ is a formula in the language of ZFC in the variables

$$x, y, z, w_1, \ldots, w_n$$

with free variables among $x, z, w_1, \ldots, w_n$ and y not free in ϕ.

4. Axiom of pairing. If x and y are sets, then there exists a set whose elements are exactly x and y.

$$\forall x \forall y \exists z \forall w (w \in z \Leftrightarrow w = x \vee w = y).$$

5. Union axiom. For any set x there is a set whose elements are exactly the elements of the elements of x:

$$\forall x \, \exists y \, \forall z (z \in y \Leftrightarrow \exists w (z \in w \wedge w \in x).$$

6. Axiom schema of collection. If ϕ is a formula in the language of ZFC with free variables among $x, y, z, w_1, \ldots, w_n$ and with a non-free variable w, then

$$\forall z \, \forall w_1, \ldots, w_n ((\forall x \in z \exists ! y \phi) \Rightarrow \exists w \forall x \in z \exists y \in w \, \phi).$$

Here $\exists ! y$ means "there exists a unique y such that...". The axiom essentially says that if $f : z \to z'$ is a function, then the image of f is set. A function $f : z \to z'$ is a triple (f, z, z') of sets, where $f \subseteq z \times z'$ and for every $x \in z$ there exists a unique $y \in z'$ with $(x, y) \in f$.

7. Axiom of infinity. The axiom essentially states that there exists a set with infinitely many members.

$$\exists x \, (\emptyset \in x \wedge \forall y (y \in x \Rightarrow y \cup \{y\} \in x)).$$

8. Axiom of power set. For any set x, there is a set y whose elements are exactly the subsets of x.

$$\forall x \exists y \forall z (z \in y \Leftrightarrow (\forall q (q \in z \Rightarrow q \in x))).$$

9. Axiom of choice. For any set X, every equivalence relation on X has a set of representatives.

This is a good point to recall in this Appendix a fact that we suppose to be known to the reader: in ZF (the Zermelo-Fraenkel set theory without the axiom of choice), the axiom of choice, the well-ordering theorem, Zorn's lemma and the existence of right inverses of surjective mappings are all equivalent conditions. We need some further terminology about partially ordered sets.

Let $(X, \leq)$ be a partially ordered set. A subset Y of X is a *chain* in X if Y, with respect to the order induced on Y by $\leq$, is a totally ordered set. An *upper bound* for a subset Z of X is an element $x \in X$ such that $z \leq x$ for every $z \in Z$.

6.1 Appendix 1: Axiomatic Set Theory

Theorem 6.1.1
The following properties are equivalent:

(a) *(Axiom of Choice) For every set I and every indexed set $\{X_i \mid i \in I\}$ of non-empty sets X_i, there exists an indexed set $\{x_i \mid i \in I\}$ such that $x_i \in X_i$ for every $i \in I$.*
(b) *For every set I and every indexed set $\{X_i \mid i \in I\}$ of non-empty sets X_i, the cartesian product $\prod_i X_i$ is non-empty.*
(c) *(Well-ordering Theorem) Every set can be well-ordered.*
(d) *(Zorn's Lemma) Let $(X, \leq)$ be a partially ordered set. Suppose that for every chain Y in X there exists an upper bound $x_Y \in X$ for Y. Then X contains at least one maximal element.*
(e) *If X and Y are sets, every surjective mapping $f \colon X \to Y$ has a right inverse, that is, there exists a mapping $g \colon Y \to X$ such that fg is the identity mapping $Y \to Y$.*

As we have said above, the equivalence of these statements is a theorem in ZF. The axiom of choice is independent from the other axioms of ZFC, and ZFC is independent from the continuum hypothesis $2^{\aleph_0} = \aleph_1$.

Grothendieck Universes.

The idea is: fix a set, which we call a *universe*, big enough because we put in it all what we need, but not too big because we want it to be a set and not a class. The formal definition is the following:

A *universe* is a set U satisfying the following properties:

(a) $X \in Y \in U \to X \in U$.
(b) $X, Y \in U \to \{X, Y\} \in U$.
(c) $X, Y \in U \to X \times Y \in U$.
(d) $X \in U \to \mathcal{P}(X) \in U$.
(e) $X \in U \to \bigcup_{Y \in X} Y \in U$.
(f) The set ω of natural numbers is an element of U.
(g) If $X \in U$ and $f \colon X \to U$ is a mapping, then $\{f(Y) \mid Y \in X\} \in U$.

Important: the axioms of ZFC do not guarantee the existence of a universe. Following Grothendieck, we adjoin a further axiom to the axioms of ZFC:

Axiom of Universes Every set is a member of a universe.
Given any universe U, if the axioms of ZFC are satisfied by the class of all sets with the relation $\in$, then they are also satisfied by the set of all sets belonging to U with the relation $\in$ between them. Hence we can argue remaining in the universe

U, which we suppose fixed once for all. In the universe, we find all what we need, and if we do not find it, we can always adjoin it to the universe thanks to the Axiom of universes. In other words, we decide to work in a set that we possibly expand.

But the problem remains: it is not possible to deal with the category **Set** of all sets in our universe, in this universe in expansion.

For the notion of class, we must introduce **NBG** (cf. page 57), the *Von Neumann-Bernays-Gödel set theory*.

Here is a list of the axioms of **NBG**. Notice that the first five coincide with five axioms of **ZFC** and deal only with sets, not classes.

1. Axiom of extensionality. Two sets are equal if they have the same elements:

$$\forall a \forall b (\forall z (z \in a \Leftrightarrow z \in b) \Rightarrow a = b).$$

2. Axiom of pairing. If x and y are sets, then there exists a set whose elements are exactly x and y.
3. Union axiom. For any set x there is a set whose elements are exactly the elements of the elements of x.
4. Axiom of power set. For any set x, there is a set y whose elements are exactly the subsets of x.
5. Axiom of infinity.

$$\exists x \, (\emptyset \in x \wedge \forall y (y \in x \Rightarrow y \cup \{y\} \in x)).$$

The remaining axioms are primarily concerned with classes rather than sets.

6. Axiom of extensionality for classes. Two classes are equal if they have the same elements:

$$A = B \Leftrightarrow \forall x (x \in A \leftrightarrow x \in B).$$

7. Axiom of foundation for classes. Every non-empty class A contains an element disjoint from A.

$$\exists x (x \in A) \Rightarrow \exists y (y \in A \wedge \neg \exists z (z \in y \wedge z \in A)).$$

Finally, the last two axioms are peculiar to **NBG**:

8. Axiom of limitation of size: For any class A, there exists a set a such that $a = A$ if and only if there is no bijection between A and the class V of all sets.
9. Class comprehension schema: For any formula ϕ containing no quantifiers over classes (it may contain class and set parameters), there exists a class A such that $\forall x (x \in A \leftrightarrow \phi(x))$.

It can be proved that **NBG** can be finitely axiomatized. What is important for us, is that in **NBG**, which is a conservative extension of **ZFC**, we can deal with classes and have the axiom of choice for classes. Thus every category has a skeleton, we have a

6.2 Appendix 2: Cardinal Numbers, Ordinal Numbers

Cardinal numbers

The problem we tackle now is measuring the size of sets. The idea is to define a mapping from the class of all sets to a class of symbols by associating to any set A a symbol called the "cardinality of A", denoted by $|A|$. Two sets A and B are *equipotent* if there exists a bijection from A onto B. We want two sets to be mapped to the same symbol if and only if they are equipotent. If a set A is finite, we can naturally associate with it the number of its elements:

$$A \mapsto n = |A|.$$

If A is *countable* (i.e., equipotent to $\mathbb{N}$) we associate with it the symbol $\aleph_0$. Call *cardinal numbers* the symbols $|A|$ associated with some set A.

Operations on cardinal numbers

Consider two cardinal numbers ξ and η, so that there exist two sets A and B with $\xi = |A|$ and $\eta = |B|$. Define

(*Addition* :) $\quad \xi + \eta = |A \mathbin{\dot\cup} B|, \quad$ where $\dot\cup$ indicates the disjoint union;

(*Multiplication* :) $\quad \xi \cdot \eta = |A \times B|$;

(*Exponentiation* :) $\quad \xi^\eta = |A^B| = |\{f: B \to A \mid f \text{ is a mapping}\}|$.

Clearly, these operations between cardinal numbers are well defined. These three binary operations have the following properties:

> **Proposition 6.2.1**
> *If ξ, η, ζ are cardinal numbers, then:*
>
> (1) $\xi + (\eta + \zeta) = (\xi + \eta) + \zeta$.
> (2) $\xi + \eta = \eta + \xi$.
> (3) $\xi(\eta + \zeta) = \xi\eta + \xi\zeta$.
> (4) $\xi(\eta\zeta) = (\xi\eta)\zeta$.
> (5) $\xi\eta = \eta\xi$.
> (6) $\xi^{\eta+\zeta} = \xi^\eta \xi^\zeta$.
> (7) $(\xi\eta)^\zeta = \xi^\zeta \eta^\zeta$.
> (8) $(\xi^\eta)^\zeta = \xi^{(\eta\zeta)}$.

These are the usual properties of addition, multiplication and exponentiation for natural numbers. Moreover, (3) and (5) imply $(\eta + \zeta)\xi = \eta\xi + \zeta\xi$.

Proof All these properties are easily checked. (1), (2) and (3) follow from the fact that we are considering disjoint unions of sets, and disjoint unions distribute with respect to cartesian product. (4) and (5) also follow from properties of cartesian product.

As far as (6) is concerned, we must show that $A^{B \dot\cup C}$ is equipotent to $A^B \times A^C$. To this end, it suffices to associate with any $f \colon B \dot\cup C \to A$ the pair of mappings $(f|_B, f|_C)$.

Similarly for (7). We must find a bijection between $(A \times B)^C$ and $A^C \times B^C$. It suffices to associate with any mapping $f \colon C \to A \times B$ the pair of mappings $(\pi_A f, \pi_B f)$.

The equality (8) is the analogue of the canonical isomorphism

$$\mathrm{Hom}(A \otimes B, C) \cong \mathrm{Hom}(A, \mathrm{Hom}(B, C)),$$

which holds for tensor product $\otimes$ of abelian groups or modules A, B, C over a commutative ring. In the category **Set** that canonical isomorphism becomes

$$\mathrm{Hom}_{\mathbf{Set}}(A \times B, C) \cong \mathrm{Hom}_{\mathbf{Set}}(A, \mathrm{Hom}_{\mathbf{Set}}(B, C)).$$

So, we must find a bijection between $C^{A \times B}$ and $(C^B)^A$. A natural bijection is given by the map $(f \colon A \times B \to C) \mapsto (F \colon A \to C^B)$, where $F \colon a \mapsto (f(a, -) \colon B \to C)$. □

Partial Order for Cardinal Numbers

In the class of cardinal numbers we can define a partial order setting $|A| \leq |B|$ if there exists an injective mapping $f \colon A \to B$. To prove that this is a partial order, reflexivity and trasitivity are very easy. For antisymmetry we need the Cantor-Schröder-Bernstein Theorem:

Theorem 6.2.2 (Cantor-Schröder-Bernstein)
Let A and B be sets such that $|A| \leq |B|$ and $|B| \leq |A|$. Then $|A| = |B|$.

In other words, if there exist injective mappings $f \colon A \to B$ and $g \colon B \to A$, then there is a bijection $h \colon A \to B$.

In order to better understand the idea behind the proof of the Cantor-Schröder-Bernstein Theorem, let us recall what we did when we showed that every permutation is a union of disjoint cycles.

6.2 Appendix 2: Cardinal Numbers, Ordinal Numbers

Consider the set $X_n = \{1, 2, \ldots, n\}$. If $f \colon X_n \to X_n$ is a bijection, we can construct its associated directed graph G_f, where the set of vertices is X_n and the set of arrows is $L = \{(i, f(i)) \mid i \in X_n\}$. So $G_f = (X_n, L)$, and G_f turns out to be a directed graph with n vertices and n arrows, for which the outdegree and the indegree at each vertex is 1 because f is bijective.

Connected directed finite graphs with outdegree and indegree equal to 1 at each vertex are only of one type: crowns on n vertices

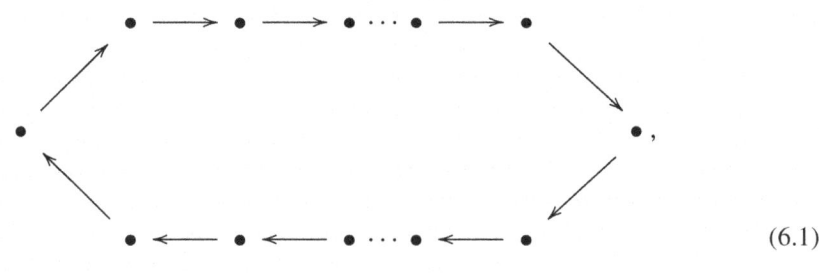

(6.1)

of which loops are a special case (crowns on 1 vertex). So the connected components of the graph G_f of a bijection f are crowns, because G_f is the disjoint union of its connected components. Therefore G_f is a disjoint union of crowns, so any permutation f is a union of disjoint cycles.

If we consider a bijection $f \colon X \to X$ where X is any set, possibly infinite, most of what we have said remains true. That is, we can construct its directed graph $G_f = (X, L)$, which has outdegree and indegree 1 in each vertex. The only difference is that the connected components of the directed graph can be infinite. In fact, a connected directed graph with outdegree and indegree 1 at every vertex is of one of these two types: either a (finite) crown (A), if there is a repetition, that is an element $x_0 \in X$ such that $f^t(x_0) = x_0$ for some $t \geq 1$, or a *line*

$$\cdots \longrightarrow \bullet \longrightarrow \bullet \longrightarrow \bullet \longrightarrow \bullet \longrightarrow \bullet \longrightarrow \cdots \qquad (6.2)$$

if there are no repetitions. Hence G_f turns out to be a disjoint union of finite crowns (A) and infinite lines (B).

Finally, if we have an injective function $f \colon X \to X$ with X any set, we can similarly build the directed graph $G_f = (X, L)$ of f, but in these case we have that the outdegree at each vertex is 1 and that the indegree at each vertex can be either 1 or 0, because f is injective. As a consequence, a connected directed graph of an injective mapping can be also a *half-line*

$$\bullet \longrightarrow \bullet \longrightarrow \bullet \longrightarrow \bullet \longrightarrow \bullet \longrightarrow \cdots \qquad (6.3)$$

Consequently, the connected components of G_f for an injective mapping f are either crowns (A), lines (B) or half-lines (C).

We are now ready for the proof of the Cantor-Schröder-Bernstein Theorem.

Proof Assume that there are injective mappings

$$f: A \to B \quad \text{and} \quad g: B \to A.$$

Consider the mapping $\Phi: X \to X$, where $X := A \,\dot\cup\, B$ is the disjoint union of A and B, and Φ is defined by

$$\Phi(x) = \begin{cases} f(x) & \text{if } x \in A \\ g(x) & \text{if } x \in B. \end{cases}$$

The mapping Φ is clearly injective. Let G_Φ be the directed graph of the mapping Φ. There is a partition of the set X of vertices into the connected components C of G_Φ: $X = \bigcup_{C \in \mathcal{C}} C$. A connected component C of G_Φ is of three possible types: a crown (A), a line (B) or a half-line (C).

Moreover, in view of how we have defined the mapping Φ, the graph G_Φ is bipartite, i. e., its arrows go either from an element of A to an element of B, or from an element of B to an element of A, but they do not connect two elements of A, nor two elements of B. So, in the first case of a crown (A) we have $|C \cap A| = |C \cap B|$, and in the other two cases of a line (B) or a half-line (C) we have $|C \cap A| = |C \cap B| = \aleph_0$.

Since $\mathcal{C}$ is a partition of $X = A \,\dot\cup\, B$, correspondingly there are a partition $\{C \cap A \mid C \in \mathcal{C}\}$ of A and a partition $\{C \cap B \mid C \in \mathcal{C}\}$ of B. Therefore, for each $C \in \mathcal{C}$, there is a bijection $\phi_C: A \cap C \to B \cap C$. Consequently, there exists a bijection $\phi: A \to B$, defined by $\phi(x) = \phi_C(x)$ for every $x \in A$, where C is the unique element of the partition $\mathcal{C}$ such that $x \in C$. □

We will not give a proof of the following proposition.

Proposition 6.2.3
Every non-empty set of cardinal numbers has a least element.

That is, every set of cardinal numbers is well-ordered by the partial order $\leq$.

Corollary 6.2.4
For every set A, B, either $|A| \leq |B|$ or $|B| \leq |A|$.

▶ **Remark 6.2.5** Partial orders are the relations that are reflexive, antisymmetric and transitive. But we could equivalently have defined them as relations that are irreflexive and transitive. Recall that a relation ρ on a set A is *irreflexive* if $a \not\rho\, a$ for every $a \in A$. We have that:

6.2 Appendix 2: Cardinal Numbers, Ordinal Numbers

> **Proposition 6.2.6**
> *Let A be a set. There is a canonical one-to-one correspondence between*
> $$\mathcal{O}(A) := \{\ \leq\ |\ \leq\ \text{is a partial order on } A\ \}$$
> *and*
> $$\mathcal{S}(A) := \{\ <\ |\ <\ \text{is a relation on } A \text{ that is irreflexive and transitive}\ \}.$$

Proof Consider the mapping $\Phi : \mathcal{O}(A) \to \mathcal{S}(A)$, where $\Phi(\leq) := \ <$ is defined by $a < a'$ if and only if $a \leq a'$ and $a \neq a'$. Conversely, let $\Psi \colon \mathcal{S}(A) \to \mathcal{O}(A)$, where $\Psi(<) := \ \leq$ is defined by $a \leq a'$ if and only if $a < a'$ or $a = a'$. It is easily checked that these two mappings are well defined and mutually inverse. □

We will keep this notation in the next pages: given a partially ordered set $(A, \leq)$, we will denote by $<$ the relation $\Phi(\leq)$ on A just defined. And, conversely, given a relation $< \ \in \mathcal{S}(A)$, we will denote by $\leq$ the relation $\Psi(<)$.

Exercise 6.2.7
Let A and B be sets and $\varphi \colon A \to B$ be an onto mapping. Show that $|B| \leq |A|$.

Exercise 6.2.8
For every set A, prove that $|\mathcal{P}(A)| = 2^{|A|}$. [*Hint:* There is a bijection $\varphi \colon \{0,1\}^A \to \mathcal{P}(A)$, defined by $\varphi(f) = f^{-1}(1)$ for every $f \in \{0,1\}^A$.

Exercise 6.2.9
For every set A, $|A| \leq |\mathcal{P}(A)|$.

Exercise 6.2.10
For every set A, show that $|A| \neq |\mathcal{P}(A)|$. [*Hint:* By contradiction, suppose there exists a bijection $\varphi \colon A \to \mathcal{P}(A)$. Set $B := \{a \in A \mid a \notin \varphi(a)\}$, so $B \in \mathcal{P}(A)$. Consequently there exists a unique $\overline{a} \in A$ such that $\varphi(\overline{a}) = B$. There are two cases: either $\overline{a} \in B$ or $\overline{a} \notin B$. If $\overline{a} \in B$, then $\overline{a} \notin \varphi(\overline{a})$, that is $\overline{a} \notin B$, which is a contradiction. In the other case $\overline{a} \notin B$, so $\overline{a} \in \varphi(\overline{a})$, that is $\overline{a} \in B$. This is also a contradiction.]

Hence, $\xi < 2^\xi$ for every cardinal ξ.

Exercise 6.2.11
Let ξ, η and ζ be cardinal numbers such that $\xi \leq \eta$. Then $\xi + \zeta \leq \eta + \zeta, \xi\zeta \leq \eta\zeta, \xi^\zeta \leq \eta^\zeta$ and $\zeta^\xi \leq \zeta^\eta$.

Recall that $\aleph_0 := |\mathbb{N}|$.

Proposition 6.2.12
Let ξ be an infinite cardinal. Then $\aleph_0 \leq \xi$.

Proof We have to show that given any infinite set A, there exists an injective mapping $f\colon \mathbb{N} \to A$. Define $f(n)$ by induction on $n \in \mathbb{N}$. Since A is infinite, we have $A \neq \emptyset$, and therefore there exists $a_0 \in A$. Set $f(0) := a_0$. Suppose that $f(0), \ldots, f(n-1)$, distinct elements of A, have been defined. Then $\{f(0), \ldots, f(n-1)\}$ is properly contained in A, whence there exists $a_n \in A \setminus \{f(0), \ldots, f(n-1)\}$. Set $f(n) := a_n$. The mapping f is injective by construction. □

Exercise 6.2.13
$2\aleph_0 = \aleph_0$. [Hint: $\mathbb{N} = \mathbb{P} \:\dot\cup\: \mathbb{D}$, where $\mathbb{P}$ is the set of even numbers and $\mathbb{D}$ is the set of odd numbers. There are two bijections $f\colon \mathbb{N} \to \mathbb{P}$, defined by $f(n) = 2n$ for every $n \in \mathbb{N}$, and $g\colon \mathbb{N} \to \mathbb{D}$, defined by $g(n) = 2n + 1$ for every $n \in \mathbb{N}$. Therefore $\aleph_0 + \aleph_0 = \aleph_0$. Finally, $\aleph_0 + \aleph_0 = (1+1)\aleph_0 = 2\aleph_0$.

Exercise 6.2.14
$\aleph_0^2 = \aleph_0$. [Hint: ($\geq$) From $1 \leq \aleph_0$, we get that $1\aleph_0 \leq \aleph_0 \aleph_0$, that is, $\aleph_0 \leq \aleph_0^2$. ($\leq$) Consider the injective mapping

$$\mathbb{N} \times \mathbb{N} \to \mathbb{N}$$

$$(a,b) \mapsto 2^a 3^b.$$

Then $\aleph_0^2 = |\mathbb{N} \times \mathbb{N}| \leq |\mathbb{N}| = \aleph_0$.]

The next lemma is a preliminary lemma necessary for the proof of Theorem 6.2.16.

Lemma 6.2.15
If ξ is an infinite cardinal and $\xi^2 = \xi$, then $\xi = 2\xi = 3\xi$.

Proof $\xi \leq 2\xi \leq 3\xi \leq \xi\xi = \xi^2 = \xi$. □

Theorem 6.2.16
If ξ is an infinite cardinal, then $\xi = \xi^2$.

6.2 Appendix 2: Cardinal Numbers, Ordinal Numbers

Proof Let X be a set of cardinality ξ, and consider

$$\mathcal{J} := \{ (E, \varphi) \mid E \subseteq X, \varphi \colon E \to E \times E \text{ a bijection} \}.$$

Since there exists a subset $E \subseteq X$ of cardinality $\aleph_0$, and $\aleph_0^2 = \aleph_0$, the set $\mathcal{J}$ is non-empty. Partially order $\mathcal{J}$ by

$$(E, \varphi) \le (E', \varphi') \text{ if } E \subseteq E' \text{ and } \varphi = \varphi'|_E^{E \times E},$$

where $\varphi'|_E^{E \times E}$ is the restriction of φ' to E with codomain $E \times E$. In other words, $(E, \varphi) \le (E', \varphi')$ if $E \subseteq E'$ and φ' is a bijection $E' \to E' \times E'$ extending φ. Then $(\mathcal{J}, \le)$ is clearly inductive, so it has a maximal element $(\overline{E}, \overline{\varphi})$ by Zorn's Lemma. Let us show that $|\overline{E}| = \xi$. This will imply that $\xi = \xi^2$, because $\overline{\varphi} \colon \overline{E} \to \overline{E} \times \overline{E}$ is a bijection, as we want to prove.

For the sake of contradiction suppose $\eta := |\overline{E}| < \xi$. Let $\zeta := |X \setminus \overline{E}|$. We claim that $\zeta > \eta$. Indeed, since X is disjoint union of $\overline{E}$ and $X \setminus \overline{E}$, we have that $\xi = \eta + \zeta$. Moreover $\eta^2 = \eta$, because $\overline{E} \in \mathcal{J}$, so $2\eta = \eta$ by Lemma 6.2.15. Thus, if $\zeta \le \eta$, then $\xi \le 2\eta = \eta$, which is a contradiction.

Hence $\zeta > \eta$, so there exists a subset $C \subseteq X \setminus \overline{E}$ of cardinality η. We will show that there is a bijective map ψ extending φ such that $(\overline{E} \cup C, \psi) \in \mathcal{J}$, contradicting the maximality of $(\overline{E}, \overline{\varphi})$. Now

$$(\overline{E} \cup C) \times (\overline{E} \cup C) = (\overline{E} \times \overline{E}) \cup (\overline{E} \times C) \cup (C \times \overline{E}) \cup (C \times C),$$

so such a map ψ exists if there is a bijection

$$C \to (\overline{E} \times C) \cup (C \times \overline{E}) \cup (C \times C).$$

But this follows from

$$|(\overline{E} \times C) \cup (C \times \overline{E}) \cup (C \times C)| = \eta^2 + \eta^2 + \eta^2 = \eta + \eta + \eta = 3\eta = \eta$$

by Lemma 6.2.15. □

From Theorem 6.2.16, we easily obtain the following corollaries.

Corollary 6.2.17
If ξ and η are non-zero cardinal numbers, one of which is infinite, then

$$\xi + \eta = \xi\eta = \max\{\xi, \eta\}.$$

Proof Without loss of generality, suppose $\xi \leq \eta$. Then

$$\eta = 0 + \eta \leq \xi + \eta \leq \eta + \eta = 2\eta = \eta$$

by Lemma 6.2.15. Moreover, by Theorem 6.2.16,

$$\eta = 1 \cdot \eta \leq \xi\eta \leq \eta^2 = \eta.$$

Thus $\eta = \xi + \eta = \xi\eta$. □

Corollary 6.2.18
If ξ is an infinite cardinal, then

$$\xi^\xi = 2^\xi.$$

Proof By Theorem 6.2.16 and the fact that $\xi < 2^\xi$, we have

$$2^\xi \leq \xi^\xi \leq (2^\xi)^\xi = 2^{\xi^2} = 2^\xi.$$

□

Corollary 6.2.19
If $1 < \xi$ and η are cardinal numbers, one of which is infinite, then

$$\xi^\eta = (\xi + \eta)^\eta = (\xi\eta)^\eta = (\max\{\xi, \eta\})^\eta.$$

Proof The case $\eta = 0$ is trivial. Assume $\eta > 0$. By Corollary 6.2.17, we have

$$\xi^\eta = (\xi + 0)^\eta \leq (\xi + \eta)^\eta = (\max\{\xi, \eta\})^\eta =$$
$$= (\xi\eta)^\eta = \xi^\eta \eta^\eta = \xi^\eta 2^\eta = (\max\{2, \xi\})^\eta = \xi^\eta.$$

□

Let us also define sums and products of infinite sets of cardinal numbers. Let $\{\xi_\lambda \mid \lambda \in \Lambda\}$ be a family of cardinal numbers and, for each $\lambda \in \Lambda$, let X_λ be a set

6.2 Appendix 2: Cardinal Numbers, Ordinal Numbers

with $|X_\lambda| = \xi_\lambda$. Assume that $X_\lambda \cap X_\mu = \emptyset$ for all $\lambda, \mu \in \Lambda$ with $\lambda \neq \mu$. Define

$$\sum_{\lambda \in \Lambda} \xi_\lambda = \left|\dot{\bigcup}_{\lambda \in \Lambda} X_\lambda\right|$$

and

$$\prod_{\lambda \in \Lambda} \xi_\lambda = \left|\prod_{\lambda \in \Lambda} X_\lambda\right| = \left|\left\{f : \Lambda \to \bigcup_{\lambda \in \Lambda} X_\lambda \mid f(\lambda) \in X_\lambda \text{ for all } \lambda \in \Lambda\right\}\right|.$$

If the sets X_λ, $\lambda \in \Lambda$, are not pairwise disjoint, we can always define their disjoint union as $\bigcup_{\lambda \in \Lambda} (X_\lambda \times \{\lambda\})$.

The axiom of choice is equivalent to the assertion that whenever X_λ is non-empty for all $\lambda \in \Lambda$, then the direct product $\prod_{\lambda \in \Lambda} X_\lambda$ is non-empty.

Exercise 6.2.20
If $\{X_\lambda \mid \lambda \in \Lambda\}$ is a family of sets such that $|X| = |X_\lambda|$ for all $\lambda \in \Lambda$, then

$$\sum_{\lambda \in \Lambda} |X_\lambda| = |X||\Lambda|.$$

Exercise 6.2.21
Show that the cardinality of $\mathbb{R}$ is $2^{\aleph_0}$. The cardinal number $c := 2^{\aleph_0}$ is sometimes called the *cardinality of the continuum*.

Exercise 6.2.22
The set $\mathscr{P}(\mathbb{N})$ of all subsets of $\mathbb{N}$ is obviously the disjoint union of $\mathscr{P}_f(\mathbb{N})$, the set of all finite subsets of $\mathbb{N}$, and $\mathscr{P}_\infty(\mathbb{N})$, the set of all infinite subsets of $\mathbb{N}$. Show that $|\mathscr{P}_f(\mathbb{N})| = \aleph_0$ and $|\mathscr{P}_\infty(\mathbb{N})| = 2^{\aleph_0}$.

Exercise 6.2.23
Show that the cardinality of the set of all continuous functions $f : \mathbb{R} \to \mathbb{R}$ is the cardinality of the continuum c. [Hint: A continuous function is completely determined by its restriction to $\mathbb{Q}$.]

Exercise 6.2.24
If V is a k-vector space, the cardinality of its dual space $V^* := \mathrm{Hom}(V_k, k)$ is $|k|^{\dim V_K}$.

Partially Ordered Sets

We now define the category **POSet** of partially ordered sets as follows. The objects of **POSet** are the partially ordered sets $(A, \leq)$. For any two partially ordered sets $(A, \leq), (B, \leq)$,

$$\mathrm{Hom}_{\mathbf{POSet}}((A, \leq), (B, \leq)) := \{f : A \to B \mid a \leq a' \Rightarrow f(a) \leq f(a')\}.$$

The morphisms of **POSet** are called *increasing maps*. A map $f: A \to B$ between linearly ordered sets is said to be *strictly increasing* if $a < a' \Rightarrow f(a) < f(a')$ for every $a, a' \in A$.

Lemma 6.2.25
Let A and B be linearly ordered sets. A mapping $f: A \to B$ is strictly increasing if and only if it is increasing and injective.

Proof If f is increasing and injective, then:

$$a < a' \Rightarrow a \leq a', a \neq a' \Rightarrow f(a) \leq f(a'), f(a) \neq f(a') \Rightarrow f(a) < f(a')$$

so f is strictly increasing. Conversely, if f is strictly increasing, then it is increasing:

$$a \leq a' \Rightarrow a < a' \vee a = a' \Rightarrow f(a) < f(a') \vee f(a) = f(a') \Rightarrow f(a) \leq f(a').$$

Finally, f is injective, because the sets are linearly ordered, hence

$$a \neq a' \Rightarrow a < a' \vee a' < a \Rightarrow f(a) < f(a') \vee f(a') < f(a) \Rightarrow f(a) \neq f(a').$$

□

The category **POSet** has the two full subcategories **LOSet** of linearly ordered sets and **WOSet** of well-ordered sets.

Lemma 6.2.26
*In **LOSet**, a mapping is an isomorphism if and only if it is a bijection and an increasing mapping.*

Proof If f is an isomorphism of linearly ordered sets, then f is a morphism, so it is increasing. Moreover, by definition of isomorphism, there exists a morphism $g: B \to A$ such that $f \circ g = \iota_B$ and $g \circ f = \iota_A$. Therefore f is a bijection. Conversely, let f be an increasing bijection. Then there exists $g: B \to A$ such that $f \circ g = \iota_B$ and $g \circ f = \iota_A$. We have to prove that g is increasing. Let $b \leq b'$. Assume by contradiction that $g(b) \not\leq g(b')$. Then $g(b') \leq g(b)$. Since f is increasing and $f \circ g = \iota_B$, this implies that $b' = f(g(b')) \leq f(g(b)) = b$, so $b' \leq b \leq b'$, hence $b = b'$. It follows that $g(b) = g(b')$, so $g(b) \leq g(b')$, a contradiction. □

6.2 Appendix 2: Cardinal Numbers, Ordinal Numbers

Ordinals

To introduce ordinals, we will follow the same idea we have made use of to define cardinal numbers: we want to associate a symbol ord(A) with every well-ordered set A in such a way that ord(A) = ord(B) if and only if A and B are isomorphic in **WOSet**.

For instance, we will set ord($\emptyset$) := 0, ord($\{1,\ldots,n\}$) := n, ord($\mathbb{N}$) := ω_0, ord($\mathbb{N} \cup \{+\infty\}$) := $\omega_0 + 1$, where the order is defined in such a way that $n < +\infty$ for all $n \in \mathbb{N}$, ord($\{n - \frac{1}{m} \mid n, m \in \mathbb{N}\}$) := ω_0^2.

> **Theorem 6.2.27**
> Let A be a well-ordered set. Let $f : A \to A$ be a strictly increasing mapping. Then $a \leq f(a)$ for every $a \in A$.

Proof Set $I := \{a \in A \mid a \not\leq f(a)\}$, so that $I := \{a \in A \mid f(a) < a\}$. We want to show that $I = \emptyset$. By contradiction, assume I non-empty. Then I has a least element $m \in I$ say, because A is well-ordered. Since $m \in I$, $f(m) < m$. Set $y := f(m)$, so $y < m$. The mapping f is strictly increasing, so $f(y) < f(m) = y$ and hence $y \in I$. But this is a contradiction, because $y < m$ and m is the least element of I. □

▶ **Remark 6.2.28**

(1) In the theorem above (Theorem 6.2.27), it is necessary that X is well-ordered. For instance, $\mathbb{R}$ is linearly ordered, but not well-ordered, and $f \colon \mathbb{R} \to \mathbb{R}$, $f(x) = x^3$ is strictly increasing. But $f(\frac{1}{2}) = \frac{1}{8} < \frac{1}{2}$.
(2) In the theorem above, it is also necessary that the function is strictly increasing. For instance, $\mathbb{N}$ is well-ordered and a constant function $f \colon \mathbb{N} \to \mathbb{N}$ is increasing, and it is not true that $x \leq f(x)$ for every $x \in \mathbb{N}$.

> **Corollary 6.2.29**
> In the category **WOSet**, a mapping $f : A \to A$ is an isomorphism if and only if $f = \iota_A$. That is, the unique increasing bijection from a well-ordered set onto itself is the identity.

Proof Clearly, ι_A is an isomorphism. Conversely, let $f \colon A \to A$ be an isomorphism. By Theorem 6.2.27, we have that $x \leq f(x)$ for every $x \in A$. Also, since f is an isomoprhism, its inverse f^{-1} is an isomorphism, so that, again, $x \leq f^{-1}(x)$ for every $x \in A$. But f is increasing, whence $f(x) \leq f(f^{-1}(x)) = x$ for every

$x \in A$. It follows that $x \leq f(x) \leq x$, so $f(x) = x$ for every $x \in A$. This proves that $f = \iota_A$. □

In this corollary it is also necessary that A is well-ordered, as the example, $f \colon \mathbb{R} \to \mathbb{R}$, defined by $f(x) = x^3$, shows. This f is an automorphism of the linearly ordered set $\mathbb{R}$.

Corollary 6.2.30
Let A and B be well-ordered isomorphic sets. Then the isomorphism $A \to B$ is unique.

Proof Let $\psi, \phi \colon A \to B$ to be two isomorphisms. The mapping $\phi^{-1} \circ \psi \colon A \to A$ is an isomorphism. By the previous corollary $\phi^{-1} \circ \psi = \iota_A$, so that $\psi = \phi$. □

Initial Segments of Well-ordered Sets

Let A be a well-ordered set. For every $a \in A$, the *initial segment of A determined by a* is the set

$$W(a) := \{x \in A \mid x < a\}.$$

Proposition 6.2.31
If A is a well-ordered set and $a \in A$, then $ord(A) \neq ord(W(a))$.

Proof By contradiction, suppose there is an isomorphism $f \colon A \to W(a) \subset A$, that is, a strictly increasing bijection. The mapping f can also be viewed as a strictly increasing injection $A \to A$ that is not onto. By Theorem 6.2.27, we have that $x \leq f(x)$ for every $x \in A$. In particular we have $a \leq f(a)$, so $f(a) \notin W(a)$, which is a contradiction. □

Proposition 6.2.32
If A is a well-ordered set, $a, b \in A$ and $a \neq b$, then $ord(W(a)) \neq ord(W(b))$.

Proof Since a well-ordered set is linearly ordered and $a \neq b$, either $a < b$ or $b < a$. For instance, assume that $a < b$. Then the initial segment of A determined

6.2 Appendix 2: Cardinal Numbers, Ordinal Numbers

by a coincides with the initial segment of $W(b)$ determined by a. Apply the previous proposition to the set $X := W(b)$ and the element $a \in X$. □

We can now define an irreflexive and transitive relation on the class of all ordinals. More precisely, if A and B are well-ordered sets, we will write $ord(A) < ord(B)$ if A is isomorphic to an initial segment of B. As usual (Proposition 6.2.6), we will also write $ord(A) \leq ord(B)$ if $ord(A) = ord(B)$ or $ord(A) < ord(B)$.

As we have already said, $<$ is an irreflexive and transitive relation, and $\leq$ is reflexive, antisymmetric and transitive.

It would be possible to prove that:

Theorem 6.2.33
Every set of ordinal numbers is well-ordered by the relation $\leq$. In particular, if A and B are well-ordered sets, then either $ord(A) < ord(B)$, or $ord(B) < ord(A)$, or $ord(A) = ord(B)$.

Notice that well-ordered sets are exactly the artinian linearly ordered sets. In fact, if a linearly ordered set is artinian, then all its non-empty subsets have minimal elements. But in a linearly ordered sets, an element is minimal if and only if it is the least element. The converse is trivial.

Formal Definition of Ordinal Numbers

Our aim is now to define a "special" *skeleton* $\mathcal{Z}$ of **WOSet**. Thus $\mathcal{Z}$ must be a class of well-ordered sets, it must be well-ordered by the relation $\leq$, and every well-ordered set X must be isomorphic to an element of $\mathcal{Z}$. To this end we will use two axioms of ZFC the Axiom of infinity and the Axiom of foundation (see the axioms of ZFC in Appendix 1.

The idea is that the elements $X \in \mathcal{Z}$ will be sets whose elements are all the elements of $\mathcal{Z}$ that precede X (i.e., $X = W(X)$ in $\mathcal{Z}$). The first elements of the well-ordered class $\mathcal{Z}$ will be, respectively, $\emptyset$, $\emptyset \cup \{\emptyset\} = \{\emptyset\}$, $\{\emptyset\} \cup \{\{\emptyset\}\} = \{\emptyset, \{\emptyset\}\}$, and so on. That is, $0 := \emptyset$, and $n + 1 := n \cup \{n\}$ for all natural numbers $n \in \mathbb{N}$.

We are ready for the formal definition of ordinal numbers.

Definition 6.2.34

An *ordinal number* is a set X with these properties:

(1) For all $x, y \in X$, either $x \in y$, or $y \in x$, or $x = y$
(2) For every $x \in X$ and every $y \in x$, one has that $y \in X$.

For instance the three sets ∅, {∅} and {∅, {∅}} constructed above have these two properties.

> **Lemma 6.2.35**
> *The relation $\in$ is irreflexive and transitive on every ordinal number.*

Proof To prove irreflexivity, use the axiom of foundation (one of the consequences of the axiom of foundation is that $A \notin A$ for every set A). Transitivity follows directly from the definition of ordinal number. □

Here are some properties of ordinal numbers:

(1) For every ordinal number X, the relation $\subseteq$ (that is $\in$ or $=$) is a well-ordering on X, and the least element of any ordinal number $X \neq \emptyset$ is $\emptyset$.

Proof The case $X = \emptyset$ is trivial. Suppose $X \neq \emptyset$, X an ordinal number. X is partially ordered by Lemma 6.2.35, and linearly ordered by Definition 6.2.34(1). In order to show that it is well-ordered, we must show that it is artinian, i.e., that there are not strictly descending chains in X. Let us prove that there are not sets $A_0, A_1, A_2, \ldots$ with $A_0 \ni A_1 \ni A_2 \ni \ldots$ Assume the contrary, and consider the set $A := \{A_0, A_1, A_2, \ldots\}$. By the Axiom of Foundation applied to the non-emptyset set A, there exists $u \in A$ such that $u \cap A = \emptyset$. Thus $u = A_n$ for some n. But $A_{n+1} \in u \cap A = \emptyset$, a contradiction. □

(2) Every element of an ordinal number is an ordinal number.
(3) If X and Y are ordinal numbers, then $X \in Y \iff X \subset Y$.

> **Theorem 6.2.36**
> *Let $\mathcal{Z}$ be the class of all ordinal numbers, partially ordered by set inclusion $\subseteq$. Then:*
>
> *(1) $(\mathcal{Z}, \subseteq)$ is well-ordered.*
> *(2) $\mathcal{Z}$ is a proper class.*
> *(3) For every $X \in \mathcal{Z}$, $X = W(X)$ in $(\mathcal{Z}, \subseteq)$.*
> *(4) $\mathcal{Z}$ is a skeleton for **WOSet**.*

Successor Ordinals, Limit Ordinals

Every ordinal number α has a *successor*, denoted by $\alpha + 1$. It is the least of the ordinal numbers $> \alpha$. More precisely, if X is an ordinal number, its successor is $X \cup \{X\}$.

Conversely, not all ordinal numbers α have an immediate predecessor, that is, the set $\{\beta \mid \beta$ is an ordinal number and $\beta < \alpha\}$ does not necessarily have a greatest element. In other words, not all ordinal numbers α are the successor $\beta + 1$ of some ordinal number β. Such ordinal numbers α are called *limit ordinals*.

Hence there are three types of ordinal numbers:

(1) The zero ordinal 0.
(2) Successor ordinals.
(3) Limit ordinals.

Arithmetic of Ordinal Numbers

There are three important binary operations on the class of ordinal numbers: addition, multiplication and exponentiation.

Addition If A and B are well-ordered sets, their disjoint union $A \cup\!\!\!\!\cdot\, B$ can be well-ordered defining, for every $x, y \in A \cup\!\!\!\!\cdot\, B$,

$$x \leq y \text{ if and only if } \begin{cases} x, y \in A \text{ and } x \leq y \text{ in } A; \\ x, y \in B \text{ and } x \leq y \text{ in } B; \\ x \in A \text{ and } y \in B. \end{cases}$$

If α, β are the ordinal numbers of A and B respectively, then $\alpha + \beta$ is by definition the ordinal number of $A \cup\!\!\!\!\cdot\, B$.

Let us list some properties of addition of ordinal numbers. For every ordinal number α, β, γ:

(1) *(Associativity:)* $\alpha + (\beta + \gamma) = (\alpha + \beta) + \gamma$.
(2) Addition of ordinal numbers is not commutative: for instance $\omega_0 + 1 \neq 1 + \omega_0 = \omega_0$. Hence addition on ordinal numbers is completely different from addition on cardinal numbers.
(3) $\alpha + 0 = 0 + \alpha = \alpha$.
(4) $\alpha < \beta \Rightarrow \gamma + \alpha < \gamma + \beta$.
(5) $\alpha < \beta \Rightarrow \alpha + \gamma \leq \beta + \gamma$.
(6) *(Left cancelativity:)* $\alpha + \beta = \alpha + \gamma \Rightarrow \beta = \gamma$.
(7) *(Subtraction:)* For every $\beta \leq \alpha$ there exists a unique γ such that $\alpha = \beta + \gamma$.
(8) Addition on ordinals is not right cancelative, as the example $1 + \omega_0 = 0 + \omega_0$ shows.

Multiplication If A and B are well-ordered sets, it is possible to endow their cartesian product $A \times B$ with the following *lexicographic order*:

$$(a, b) \leq (a', b') \text{ if and only if either } b \leq b' \text{ and } b \neq b', \text{ or } b = b' \text{ and } a \leq a'.$$

Correspondingly, one defines multiplication of ordinal numbers.

For instance:

Example 6.2.37
$\omega_0 \cdot 2 = \omega_0 + \omega_0$ and $2 \cdot \omega_0 = \omega_0$. Hence multiplication of ordinal numbers is not commutative.

Some properties of multiplication of ordinal numbers: for every ordinal number α, β, γ

(1) *(Associativity:)* $\alpha(\beta\gamma) = (\alpha\beta)\gamma$.
(2) $\alpha \cdot 0 = 0 \cdot \alpha = 0$.
(3) $\alpha \cdot 1 = 1 \cdot \alpha = \alpha$.
(4) *(Distributivity on the left:)* $\alpha(\beta + \gamma) = \alpha\beta + \alpha\gamma$.
(5) Distributivity on the right does not hold, as Example 6.2.37 shows: $(\beta + \gamma)\alpha \neq \beta\alpha + \gamma\alpha$ for $\beta = \gamma = 1$ and $\alpha = \omega_0$.

Exponentiation

For every α, β, the ordinal number α^β is defined by *transfinite induction* on β via the equalities:

$$\begin{cases} \alpha^0 = 1, \\ \alpha^{\beta+1} = \alpha^\beta \cdot \alpha, \\ \text{if } \beta \text{ is a limit ordinal, } \alpha^\beta = \sup\{\alpha^\delta \mid \delta < \beta\}. \end{cases}$$

Some properties of exponentiation:

(1) $\alpha^0 = 1$.
(2) $0 < \alpha \Rightarrow 0^\alpha = 0$.
(3) $1^\alpha = 1$.
(4) $\alpha^1 = \alpha$.
(5) $\alpha^\beta \cdot \alpha^\gamma = \alpha^{\beta+\gamma}$.
(6) $(\alpha^\beta)^\gamma = \alpha^{\beta \cdot \gamma}$.

We conclude with some further remarks.

6.2 Appendix 2: Cardinal Numbers, Ordinal Numbers

(1) *Cantor normal form:* Every ordinal number α can be written in a unique way in the form $\omega_0^{\beta_1} c_1 + \omega_0^{\beta_2} c_2 + \cdots + \omega_0^{\beta_k} c_k$, where k is a natural number, $c_1, c_2, \ldots, c_k$ are positive integers, and $\beta_1 > \beta_2 > \cdots > \beta_k \geq 0$ are ordinal numbers.

This is exactly the way we write non-negative integers in basis 10, or more generally in basis b for every $b \geq 2$: every non-negative integer n can be written is a unique way in the form $10^{\beta_1} c_1 + 10^{\beta_2} c_2 + \cdots + 10^{\beta_k} c_k$, where k is a natural number, $c_1, c_2, \ldots, c_k$ are positive integers < 10, and $\beta_1 > \beta_2 > \cdots > \beta_k$ are natural numbers. In some sense, writing ordinals in Cantor normal form corresponds to writing them "in basis ω_0".

Writing ordinal numbers in their Cantor normal form allows us to see that ordinal numbers form an additive free commutative monoid on the set $\{ \omega_0^\alpha \mid \alpha$ an ordinal number $\}$. Addition and multiplication between ordinal numbers in their Cantor normal form is very easy. Given an ordinal number $\alpha = 10^{\beta_1} c_1 + 10^{\beta_2} c_2 + \cdots + 10^{\beta_k} c_k$ in Cantor normal form, then α is the zero ordinal number if $k = 0$; the ordinal α is a successor ordinal if $k > 0$ and $\beta_k = 0$; and α is a limit ordinal if $k > 0$ and $\beta_k > 0$.

(2) Ordinal numbers form a seminearring. A *semiring* is an algebraic structure $(S, +, \cdot)$ with the same binary operations $+$ and $\cdot$ and satisfying the same axioms as a ring, but for which $(S,)$ is not necessarily an abelian group, but only a commutative monoid. The main example of semiring is $(\mathbb{N}, +, \cdot)$. *Nearring* means a ring which respect to addition is a group that is not necessarily abelian, and with only one distributivity and not the other. The main example of nearring is $(G^G, +, \circ)$, where G is an additive (not necessarily abelian) group, G^G is the set of all mappings $G \to G$, $+$ is pointwise addition, and $\circ$ is composition of mappings. Cardinal numbers form a semiring.

(3) There is a theory of factorization of ordinal numbers as a product of primes. See for instance https://en.wikipedia.org/wiki/Ordinal_arithmetic>Factorization_into_primes.

(4) Cardinal numbers can now be formally defined as particular ordinal numbers. A cardinal $\aleph$ is identified with the least ordinal number of cardinality $\aleph$. That is, a cardinal number α is an ordinal number α that is not equipotent to any ordinal $\beta < \alpha$.

For Further Study...

As recommended bibliography for further study, I suggest to read some classics. In particular I suggest the following books:

For Module Theory, the book [1] by Anderson and Fuller, and the books [18] and [19] by Tsit Yuen Lam. The book by Anderson and Fuller is more focused on the structure of modules, the books by Lam on the structure of rings as well.

For Category Theory, the book [21] by Saunders Mac Lane.

For Homological Algebra, the book [24] by Joseph Rotman.

For Representation Theory, the book [26] by Simson, Skowronski and Assem.

For ordinal numbers and cardinal numbers, the second chapter of the book "Topology" [7] by James Dugundji.

To conclude, I would like to cite some very good textbooks of Algebra, though the topics they cover are not strictly related to the topic we have treated in the present textbook: the book [12] by Garling, of Field Theory and Galois Theory; the book [2] by Atiyah and Macdonald, of Commutative Algebra (commutative rings and modules over commutative rings); my two books [9] and [10], about direct-sum decompositions of modules; the book [5] of Universal Algebra; and finally the two books [13] and [14], of "Basic Algebra", by Jacobson.

Bibliography

1. F.W. Anderson, K.R. Fuller, *Rings and Categories of Modules*, 2nd edn. (Springer-Verlag, New York, 1992)
2. M.F. Atiyah, I.G. Macdonald, *Introduction to Commutative Algebra* (Addison-Wesley Publishing Co., Reading, 1969)
3. B. Baumslag, A simple way of proving the Jordan-Hölder-Schreier theorem. Am. Math. Monthly **113**(10), 933–935 (2006)
4. G.M. Bergman, Every module is an inverse limit of injectives. Proc. Am. Math. Soc. **141**, 1177–1183 (2013)
5. S. Burris, H.P. Sankappanavar, *A Course in Universal Algebra*. Graduate Texts in Mathematics, vol. 78 (Springer-Verlag, New York-Berlin, 1981). The "Millennium Edition" is available online at https://www.math.uwaterloo.ca/snburris/htdocs/UALG/univ-algebra2012.pdf
6. P.M. Cohn, Hereditary local rings. Nagoya Math. J. **27**, 223–230 (1966)
7. J. Dugundji, *Topology*. Allyn and Bacon Series in Advanced Mathematics (Allyn and Bacon, Inc., Boston-London-Sydney, 1978)
8. S. Eilenberg, S. MacLane, General theory of natural equivalences. Trans. Am. Math. Soc. **58**, 231–294 (1945)
9. A. Facchini, *Module Theory. Endomorphism Rings and Direct Sum Decompositions in Some Classes of Modules*. Progress in Mathematics, vol. 167 (Birkhäuser Verlag, Basel, 1998). Reprinted in Modern Birkhäuser Classics, Birkhäuser/Springer Basel AG, Basel, 2012
10. A. Facchini, Semilocal Categories and Modules with Semilocal Endomorphism Rings. Progress in Mathematics, vol. 331 (Birkhäuser/Springer, Cham 2019)
11. F.G. Frobenius, Über lineare Substitutionen und bilineare Formen. J. Reine Angew. Math. **84**, 1–63 (1878)
12. D.J.H. Garling, *A Course in Galois Theory* (Cambridge University Press, Cambridge, 1986)
13. N. Jacobson, *Basic Algebra. I* (W. H. Freeman and Company, New York, 1985)
14. N. Jacobson, *Basic Algebra. II* (W. H. Freeman and Company, New York, 1989)
15. A.V. Jategaonkar, A counter-example in ring theory and homological algebra. J. Algebra **12**, 418–440 (1969)
16. I. Kaplansky, Projective modules. Ann. Math (2) **68**, 372–377 (1958)
17. I. Kaplansky, On the dimension of modules and algebras, X, A right hereditary ring which is not left hereditary. Nagoya Math. J. **13**, 85–88 (1958)
18. T.Y. Lam, *Lectures on Modules and Rings*. Graduate Texts in Mathematics, vol. 189 (Springer-Verlag, New York, 1999)
19. T.Y. Lam, *A First Course in Noncommutative Rings*. Graduate Texts in Mathematics, vol. 131 (Springer-Verlag, New York, 2001)
20. D. Lazard, Sur les modules plats. C. R. Acad. Sci. Paris **258**, 6313–6316 (1964)
21. S. Mac Lane, *Categories for the Working Mathematician*, 2nd edn. (Springer-Verlag, New York-Berlin, 1997)

22. J.H. Maclagan-Wedderburn, A theorem on finite algebras. Trans. Am. Math. Soc. **6**, 349–352 (1905)
23. D.S. Passman, *A Course in Ring Theory* (AMS Chelsea Publishing, Providence, 2004)
24. J.J. Rotman, An Introduction to Homological Algebra. Pure and Applied Mathematics, vol. 85 (Academic Press Inc./Harcourt Brace Jovanovich Publishers, New York/London, 1979)
25. A.H. Schofield, Artin's problem for skew field extensions. Math. Proc. Camb. Philos. Soc. **97**, 1–6 (1985)
26. D. Simson, A. Skowronski, I. Assem, *Elements of the Representation Theory of Associative Algebras* (Cambridge University Press, Cambridge, 2007)
27. L.W. Small, An example in Noetherian rings. Proc. Nat. Acad. Sci. USA **54**, 1035–1036 (1965)
28. L.W. Small, Hereditary rings. Proc. Nat. Acad. Sci. USA **55**, 25–27 (1966)

Index

A
Addition of ordinal numbers, 211
Algebra, 53
Anti-isomorphism, 14
Artinian module, 87
Artinian partially ordered set, 87
Augmentation map, 127
Automorphism, 15
Axiom of choice, 195

B
Bimodule, 19

C
Cantor normal form, 212
Cardinality of the continuum, 205
Cardinal number, 197
Categories, 7
    equivalent, 55
    of pointed sets, 9
    small, 10
Center of a ring, 6
Chain, 194
Chain complex of modules, 61
Complex of modules, 61
Composition (in a category), 8
Composition factors of a module, 96
Composition length of a module, 96
Composition series, 93
Concatenation of words, 51
Congruence in a group, 23
Congruence in a module, 24
Congruence in a ring, 24
Congruence on a semigroup, 22
Countable set, 197

D
Dedekind domain, 77
Dimension of a right vector space, 17
Direct product of rings, 109
Direct product of rings, internal, 109
Divisible abelian group, 139
Division ring, 5
Dual category, 8
Duality, 55

E
Empty mapping, 9
Endomorphism, 15
Endomorphism ring, 5, 17
Epimorphism, 15
Equalizer, 184
Equipotent sets, 197
Equivalence, category, 55
Equivalent series, 93
Essential, element in a lattice, 156
Exact sequence, 61
Exponentiation of ordinal numbers, 212

F
Factors of a series, 93
Fitting's Lemma, 133
Forgetful functor, 11
Free set of generators, in a module, 41, 51
Functor, 10
    contravariant, 11
    covariant, 10
    dense, 54
    essentially surjective, 54
    faithful, 54
    full, 54
    fully faithful, 54
Functorial isomorphism, 26

G
Goldie dimension, 154, 156
Group, 23

I
Ideal, proper, 3
Ideal, right, left, two-sided, 2
Idempotent, 78
Infinite Goldie dimension, 154, 156
Initial segment, 208
Injective envelope, 144, 146
Integral domain, 4
Interval, 155
Invariant basis number (IBN), 59
Inverse of an element in a group, 23
Invertible element, 4
Irreflexive relation, 200
Isomorphism, 15
Isomorphism between categories, 12
Isomorphism class, 81
Isomorphism, in a category, 25
Isomorphism of functors, 26
Isomorphism theorems, 34

J
Join-independent, 155

K
K-algebra, 53
Kernel, 3
Kernel of a semigroup morphism, 23
Krull-Schmidt-Remak-Azumaya Theorem, 135

L
Leavitt algebra, 60, 61
Left vector space, 13
Length of a module, 96
Length of a series, 93
Limit ordinals, 211
Linear mapping, 4
Linear transformation, 4

M
Maschke's theorem, 125
Modular identity, 155
Module, free, 41
Module, indecomposable, 131
Module, injective, 137
Module, left, 12
Module morphism, 13
Module, right, 14
Module, semisimple, 84
Module, simple, 82
Modules of finite composition length, 96
Modules of finite length, 96
Monoid, 23
Monoid of words, 51
Monomorphism, 15
Morphism, 3
Morphism (in a category), 7
Multiplication of ordinal numbers, 212

N
Nakayama's Lemma, 118
Natural isomorphism of functors, 26
Natural transformation, 25
Nil ideal, 78
Nilpotent, 78
Noetherian module, 87
Noetherian partially ordered set, 87

O
Objects of a category, 7
Opposite category, 8
Opposite ring, 14
Ordinal number, 209
Orthogonal idempotents, 110

P
Preordered set, 8
Principal ideal domain (PID), 76
Product category, 59
Projective cover, 143
Projective module, 64
Proper refinement of a series, 93

Q
Quaternions, 7
Quotient ring, 3

R
Rank of a free module, 47
Refinement of a series, 93

Index

Right artinian ring, 88
Right noetherian ring, 88
Ring anti-homomorphism, 14
Ring, commutative, 1
Ring, hereditary, 76
Ring, local, 129
Ring, perfect, 144
Ring, semisimple artinian, 102
Ring, simple, 3

S

Scalar multiplication, left, 12
Scalar multiplication, right, 14
Schur's Lemma, 83
Semigroup, 22
Semigroup ring, 126
Sequence of modules, 61
Series, 93
Set of generators, in a module, 51
Skeleton, 57
Socle, 85
Split exact sequence, 64
Subcategory, 56
Subcategory, full, 56
Sub-division ring, 17
Subgroup, normal, 23
Submodule, 22
    essential, 144
    large, 144
Subring, 2
Subsemigroup, 22
Successor ordinal, 211

T

Tensor product, 68
Theorem P, 144
T-nilpotent, 144
Transposition, 14

U

Uniform, 156
Universal property, adjoining an identity, 3
Universal property of direct product, 32
Universal property of direct sum, 32
Universal property of free modules, 43
Universal property of free monoids, 51
Universal property of tensor product, 68
Upper bound, 194

V

Von Neumann-Bernays-Gödel set theory, 58

W

Well-order, 76
Well-ordering theorem, 195
Word, 51

Z

Zermelo-Fraenkel set theory, 57
Zero-divisor, 4
Zero sequence, 0-sequence, 61
Zorn's lemma, 195

The manufacturer's authorised representative in the EU is Springer Nature Customer Service Centre GmbH, Europaplatz 3, 69115 Heidelberg, Germany. If you have any concerns regarding our products, please contact ProductSafety@springernature.com

Printed and bound by CPI Group (UK) Ltd, Croydon, CR0 4YY

26/03/2026

02078943-0009